NONPARAMETRIC STATISTICS

THEORY AND METHODS

NONPARAMETRIC STATISTICS

THEORY AND METHODS

Jayant V Deshpande
University of Pune, India

Uttara Naik-Nimbalkar
University of Pune, India

Isha Dewan
Indian Statistical Institute, India

World Scientific

NEW JERSEY · LONDON · SINGAPORE · BEIJING · SHANGHAI · HONG KONG · TAIPEI · CHENNAI · TOKYO

Published by

World Scientific Publishing Co. Pte. Ltd.

5 Toh Tuck Link, Singapore 596224

USA office: 27 Warren Street, Suite 401-402, Hackensack, NJ 07601

UK office: 57 Shelton Street, Covent Garden, London WC2H 9HE

Library of Congress Cataloging-in-Publication Data

Names: Deshpande, J. V., author. | Naik-Nimbalkar, Uttara, author. | Dewan, Isha, author.
Title: Nonparametric statistics : theory and methods / by Jayant V. Deshpande
    (University of Pune, India), Uttara Naik-Nimbalkar (University of Pune, India),
    Isha Dewan (Indian Statistical Institute, India).
Description: New Jersey : World Scientific, 2017. | Includes bibliographical references and index.
Identifiers: LCCN 2017029415 | ISBN 9789814663571 (hardcover : alk. paper)
Subjects: LCSH: Nonparametric statistics. | Distribution (Probability theory)
Classification: LCC QA278.8 .D37 2017 | DDC 519.5/4--dc23
LC record available at https://lccn.loc.gov/2017029415

British Library Cataloguing-in-Publication Data

A catalogue record for this book is available from the British Library.

CONTENTS

Chapter 1

PRINCIPLES OF STATISTICAL INFERENCE

1.1 Introduction

Statistical inference is the methodology to discover statistical laws regarding the outcomes of random experiments and phenomena. Randomness is prevalent in many aspects of everyday experience. Whether it will rain or not on a particular day, the price of a particular stock on the stock market, the outcome of the toss of a coin, the weight of a newly born baby, the increase in milk yield of a cow by adding certain supplements to its feed, the time required for a chemical reaction to be completed in the presence of a fixed level of a catalytic agent, the number of emissions from a radio-active source in a fixed interval of time, the time for which an electric bulb will function before it fuses, are all unpredictable prior to the event. Some of the above may be regarded as outcomes of controlled experiments whereas others are observations on uncontrolled phenomena. However the common theme of unpredictability, i.e., randomness of outcomes runs through all of them. It is impossible to exactly predict the outcome of the next trial of any of the above or similar cases.

All the same, it is our common experience that these outcomes of random trials exhibit certain patterns of regularity in long sequences. These patterns of regularity are called statistical laws governing the outcomes. For example, if the same coin is tossed again and again, the ratio of number of heads to the total number of tosses seems to stabilize to a number p $(0 \leq p \leq 1)$ as the number of tosses increases indefinitely. We call the coin fair or unbiased if p happens to be $1/2$. This number p, which may be thought to be intrinsic to the coin in question, is called the probability of obtaining a head at a trial and may be said to be the statistical law governing the outcomes of the random experiment consisting of tossing this coin. In general, the purpose of statistical inference is to discover the statistical laws governing the outcomes of a random experiment by repeated performance of the experiment or by making repeated observations of a phenomenon.

1.2   Mathematical Structure on Random Experiments

In order to make progress in the analysis of the situation we have to introduce and impose certain useful mathematical structure on the random experiment being conducted or the random phenomena being observed.

For any random experiment under study we assume that the set of all possible outcomes is well defined and known. For example, the set for tossing a coin consists of two outcomes only: $\{H, T\}$, head and tail; the set for counting the number of emissions from radio active source is $\{0, 1, 2, \cdots\}$ the class of all non-negative integers without a finite upper bound, the set for the life of an electric bulb is half of the real line $[0, \infty)$, again without a finite upper bound. This set of all possible outcomes is called the sample space of the random experiment. Sometimes, for the sake of convenience the sample space assumed for a random experiment might be somewhat different from strict reality; in the coin tossing experiment standing on edge, losing the coin, etc. are not taken into account. Also with measuring quantities like weight, length, time, etc. the actual observations are accurate only to the extent of the smallest possible unit that can be recorded by the measuring instrument, say, one gram, one millimetre, one second, etc. But again for the sake of convenience continuous intervals are accepted as the sample spaces. Such simplifications are often carried out wherever no ambiguity is involved.

We usually denote the sample space of a random experiment by the Greek letter Ω, the elements in it, i.e., the individual outcomes (or elementary events) by w and subsets of it, called events, by A, B, etc. We need not and do not always take into consideration all possible subsets of Ω, but only a subclass of these is often relevant and sufficient. The subclass of subsets which is to be taken under consideration must, however, satisfy the characteristic properties of a σ - algebra. These may be specified as: $\mathbf{A}$ is a σ - algebra of subsets of Ω if

(i) $\Omega \in \mathbf{A}$,

(ii) If $A_1, A_2, \cdots, \in \mathbf{A}$, then $\cup_{i=1}^{\infty} A_i \in \mathbf{A}$, and

(iii) If $A \in \mathbf{A}$ then $A^c \in \mathbf{A}$.

PREFACE

Books on Nonparametric Statistics are not as numerous as, say, those on Design of Experiments, or Regression Analysis. We feel that there is a need for an alternative text book for students which can also be a reference book for practitioners of statistical methods. The present book is offered with this need in view. The emphasis here is on the heuristic and theoretical base of the subject along with the usefulness of Nonparametric Methods in various situations. The audience we have in mind is that of advanced undergraduate and graduate students along with users of these methods.

The first chapter is an Introduction to Statistical Inference in general with the role of Nonparametric Statistics within it. The second to ninth chapters deal with classical methods, and the last three chapters deal with more computation intensive methods which are often only asymptotically nonparametric. The book ends with an Appendix which brings together many of the probabilistic results required to prove the asymptotic distribution theory, relative efficiency, etc. We also include some examples to illustrate the methodology and some exercises for the students. We have not included any tables of critical points as they are now generally available in common software packages along with programs to calculate the statistics. At many places we advocate the use of the public domain software package R.

We have extensive experience in teaching such courses, in developing such methodology and also applying it in practice. We hope that the road map we provide here is effective towards these three aims.

Prof R.V.Ramamoorthy read an earlier version of the chapter on Bayesian Nonparametric Methods and we gratefully acknowledge his comments which were useful in improving the chapter.

We thank our respective families for their support during this work and the authorities of the various institutes where we worked in the last few years.

Jayant V. Deshpande
University of Pune
Uttara V. Naik-Nimbalkar
University of Pune & now at Indian Institute of Science Education and Research, Pune
Isha Dewan
Indian Statistical Institute, New Delhi

If $\Omega = \{H, T\}$, then $\mathbf{A} = \{\phi, \{H\}, \{T\}, \{H, T\}\}$ is the relevant σ - algebra. If $\Omega = \{0, 1, 2, \cdots\}$ then the relevant σ algebra would be the class consisting of all elementary events $\{0\}, \{1\}, \{2\}, \cdots$, their countable unions and complements which is the power set of Ω. If the random experiment consists of measuring an angle, then the sample space Ω would be $(0, 2\pi]$ with a possible σ - algebra given by all half open intervals of the type $(a, b]$ with $0 < a \le b \le 2\pi$, their complements and countable unions, which leads to the Borel σ - algebra of subsets of Ω.

Over the mathematical structure provided by $(\Omega, \mathbf{A})$, we then define the probabilistic or statistical law governing the occurrence of the various outcomes of the random experiment. It is a probability measure defined by the set function P from $\mathbf{A}$ to $[0, 1]$ and is assumed to satisfy the following axioms

(i) $P(\Omega) = 1$.
(ii) $P(\cup_{i=1}^{\infty} A_i) = \sum_{i=1}^{\infty} P(A_i)$, provided $A_i \cap A_j = \phi \ \forall \ i \ne j$, (countable additivity).

The measure $P(A)$ of any event A is said to be its probability. Loosely speaking it is the limiting value of the ratio of trials with outcomes belonging to A to the total number of trials, as the number of trials tends to infinity. If a fair coin is tossed then one may specify $P\{H\} = \frac{1}{2} = P\{T\}$ i.e., equal probabilities for events $\{H\}$ and $\{T\}$. Then $P(\Omega) = 1$ and $P(\phi) = 0$ completes the distribution of the probabilities on the entire σ - algebra. On the other hand if it is not known to be a fair coin then all possible probability distributions are specified by $P\{H\} = p$, $0 \le p \le 1$. Then $P\{T\} = 1 - p$, $P(\Omega) = 1$ and $P(\phi) = 0$, completes the probability distribution. The above description would hold for any random experiment whose sample space contains only two outcomes, e.g., {Rain, No rain}, { Defective, Nondefective}, etc. More complex experiments will have larger sample spaces and would require more sophisticated probability measures over the corresponding σ - algebras of events. Building upon the foundations provided by Laplace, Bernoulli, Gauss and others, an extremely fruitful and elegant theory of Probability has been developed in the twentieth century, with the above axiomatization first given by Kolmogorov. Outstanding textbooks by [Cramér (1974)], [Feller (1968)], [Feller (1971)], [Loève (1963)], [Fisz (1963)], [Billingsley (1995)] and others cover the relevant material more than adequately.

We shall borrow freely from the results of probability theory as long as it is thought that they are well known. Sometimes, results which are purely probabilistic may be discussed in the Appendix because of their importance in the theory of nonparametric inference.

1.3   Random Variables and their Probability Distributions

The triplet $(\Omega, \mathbf{A}, \mathbf{P})$ consisting of the sample space, the relevant σ - algebra of its subsets and the probability measure on it completely determines the probabilistic or statistical law governing the outcomes of a random experiment. The first two of these, viz., Ω and $\mathbf{A}$ are inherent in the structure of the experiment and are always taken as known. If P, the true probability measure is also known then there is no task left for the statistician, because one can then specify the probability of any event belonging to $\mathbf{A}$. However, when the probability distribution P is unknown then the statistician endeavours either to completely determine it or to answer various questions regarding it on the basis of the outcomes recorded in several repetitions of the experiment. In this book we shall assume that the repetitions (replications) of the experiment obey the same law and are statistically independent, that is, the outcomes of the earlier replications do not have any bearing on the performance of the later replications. It is possible to analyze experiments for which the above conditions do not hold, but we will not discuss those techniques.

The examples of random experiments cited at the beginning bring out the fact that the outcomes of these may not always be numerical. Sometimes they may be qualitative. Our development is going to vitally depend upon the numerical (or at least ordinal) properties of the outcomes. Therefore we convert the outcomes to numerical entities through the device of a random variable. It is a function $X(w)$ on Ω to a convenient subset $\mathbf{X}$ of the real line R

$$X(w) : \Omega \to \mathbf{X} \subset R.$$

Through the function X, every elementary outcome w is associated with a number $X(w)$. This number $X(w)$ is itself regarded as the outcome of the experiment for subsequent use. The condition on a function X from $\Omega \to R$ for it to be regarded as a random variable, is that it should be measurable with respect to the σ - algebra $\mathbf{A}$. For our purpose it is adequate to see that for any half open interval of the type $(-\infty, a]$, with $-\infty < a < \infty$, of the real line, its inverse image $X^{-1}(-\infty, a] = \{w : -\infty < X(w) \le a\}$,

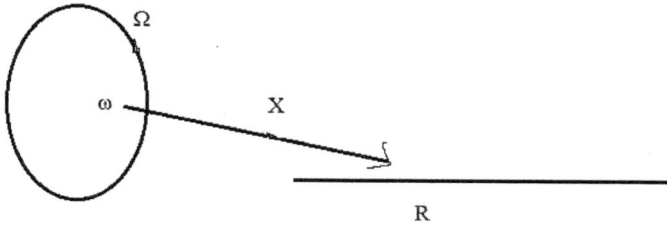

Fig. 1.1 Random variable

which is a subset of Ω, be a member of the chosen σ - algebra **A**. Then the probability under P of the set $X^{-1}(-\infty, a]$ is attributed to the subset $(-\infty, a]$ of the real line. In short we formulate the probability distribution of the random variable X through the following correspondence

$$P[-\infty < X \le a] = P_X\{(-\infty, a]\} = P\{X^{-1}(-\infty, a]\}.$$

It is often denoted by a function $F(a)$, i.e. $F(a) = P[-\infty < X \le a]$, called the cumulative distribution function, c.d.f. for short, of the random variable X. This function carries all the probabilistic information regarding the random variable X. It is known to satisfy the following properties:

(i) $\lim_{x \to -\infty} F(x) = 0$, $\lim_{x \to \infty} F(x) = 1$.
(ii) It is nondecreasing.
(iii) It is right continuous.

If the random experiment consists of tossing a coin with probability of showing head as p, then one may associate a random variable $X(H) = 1$, $X(T) = 0$, with it which will then have the c.d.f.

$$F(x) = \begin{cases} 0 & \text{if } -\infty < x < 0 \\ 1 - p & \text{if } 0 \le x < 1 \\ 1 & \text{if } 1 \le x < \infty. \end{cases}$$

The function $F(x)$ changes with values of p ranging between $[0, 1]$. Thus, unless we know what p is, we have a whole family of c.d.f.'s which may be denoted by $\mathbf{F} = \{F_p, 0 \le p \le 1\}$.

At this point we shall take for granted familiarity with properties of various c.d.f.'s and the corresponding probability mass functions (p.m.f.), probability density functions (p.d.f.), continuity with respect to a suitable

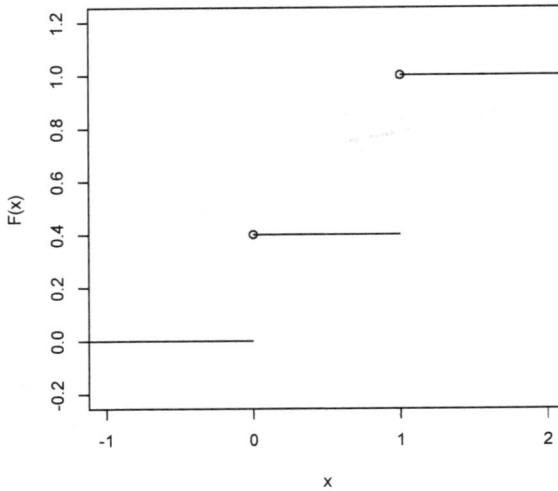

Fig. 1.2   Distribution function

measure (often the Lebesgue measure), their Radon - Nikodym derivatives, moments, moment generating functions (m.g.f.), characteristic functions (c.f.) and other entities involved in the calculus of probabilities. Development up to the standard laws of large numbers and central limit theorems will also be expected to be known. A few results have been listed in the Appendix.

1.4   Parametric and Nonparametric Statistical Models

Usually with certain introspection and analysis of the experiment we reach a family of probability distributions for the outcomes, or often a family of c.d.f.'s for the random variable. Thus our original triplet $(\Omega, \mathbf{A}, \mathbf{P})$ is now replaced with a new one $(\mathbf{X}, \mathbf{B}, \mathbf{F})$ where $\mathbf{X}$ is the range or the sample space of the random variable, $\mathbf{B}$ is the Borel σ - algebra, i.e. the σ - algebra of subsets of R generated by half open intervals of the type $(-\infty, x]$ by completing it under countable unions and complements, and a family $\mathbf{F}$ of c.d.f.'s in which the true probability law of the outcomes is known to lie.

(i)   In the single coin tossing experiment we have above specified the family of c.d.f.s.

(ii)   If a coin is tossed n times and the random variable gives the number of heads in these n trials without regard to the order of their occurrence then the Binomial family $B(n, p), 0 \leq p \leq 1$, will be appropriate.

(iii) For electron emissions it is found that the Poisson family $P(\lambda)$ with parameter $\lambda > 0$ is applicable.

(iv) For many experiments the Normal (Gaussian) family $N(\mu, \sigma^2)$ with $-\infty < \mu < \infty$ and $0 < \sigma^2 < \infty$ is found suitable.

(v) Sometime all that we may be able to say about the random variable is that its probability distribution is continuous and symmetric about a centre μ, i.e., $P[X \leq \mu - x] = P[X \geq \mu + x] \ \forall \ x \geq 0$ or equivalently

$$F(\mu - x) = 1 - F(\mu + x-) \ \forall \ x > 0.$$

(vi) An even less informative family (hence larger) of probability distributions is the family of continuous distributions, i.e. $\mathbf{F} = \{$all F such that it does not have any jumps$\}$.

(vii) Another family is that of all c.d.f.'s with expectation existing i.e. $\mathbf{F} = \{$all F such that $\int_{-\infty}^{\infty} x dF(x) < \infty\}$ where the integral may be interpreted as a Lebesgue - Stieltjes integral without worrying whether F is continuous or not.

Such a family chosen on the basis of prior information, without any recourse to the data obtained through the conduct of experiments, is called the statistical model for the experiment. Choice of the model can be data based, but we shall not deal with such families.

In the first four of the above families the functional form of the c.d.f., or equivalently that of the p.m.f. or p.d.f., is known. A specific member within the family can be pinpointed by choosing the value of p in $[0, 1]$ in cases (i) and (ii), the value of λ in $(0, \infty)$ in case (iii) and the value of (μ, σ^2) in $(-\infty, \infty) \times (0, \infty)$ in case (iv). Since in the remaining three cases the model does not specify the functional form of the c.d.f., in order to choose a single member from it first we will have to agree on the functional form and then specify the member within it by assigning the values of the constants appearing in it. The constants like $p, \lambda, (\mu, \sigma^2)$ which appear in the functional form of the c.d.f. are called the parameters of the distribution.

Thus we are able to classify statistical models as

(i) parametric: those in which the functional form of the c.d.f. is known and only the values of at most a finite number of real valued parameters may remain to be specified and

(ii) nonparametric: those in which the functional form of the c.d.f. is yet to be specified.

The same classification may equivalently be described as

(i) Parametric model: a model in which the members can be put into 1:1 correspondence with some subset of a finite dimensional Euclidean space,

(ii) Nonparametric model: a model in which the members can not be put into 1:1 correspondence with any subset of a finite dimensional subset of the Euclidean space.

This classification may not be totally devoid of ambiguity, but is serviceable in most contexts.

Once a model has been adopted the procedures of statistical inference are invoked to answer further questions about the true probability distribution. In parametric models the questions are then about the values of the parameters which index the probability distributions within the specified family. In nonparametric models the questions could be about certain functionals of the c.d.f. like mean, median, other quantiles, etc. or regarding qualitative aspects of c.d.f., e.g., is the c.d.f. symmetric, is one c.d.f. $F(x)$ less than or equal to another c.d.f. $G(x)$ for all x? etc. The methods of inference developed for parametric models tend to depend heavily upon the parametric family under consideration. Methods developed as 'optimal' for one family may not be so for another. Whereas methods developed for nonparametric families may be 'reasonable' for the whole model consisting of many parametric families without being 'optimal' for any. While laying down 'optimality', 'goodness' or 'reasonableness' criteria, we shall primarily adhere to the so called frequentist principles of statistical inference as developed by the K. Pearson - Fisher - Neyman - E. S. Pearson school and widely adopted by the statistical community. Criteria based on subjective prior distributions (Bayesian) in the context of nonparametric models are introduced in Chapter 12.

1.5   Estimation of Parameters

The purpose of statistical inference is to discover the statistical law governing a random variable within the specified model, or advance our knowledge regarding this law by further restricting the model with the help of observations. In a parametric model the point estimation procedure uses the observations to estimate the value of all or some of the unknown parameters of the given parametric family. The estimated value is accompanied with its standard error which is the standard deviation of the sampling distributions of the statistic (function of the observations) that is used as the estimator, to provide an idea of the possible spread of such estimates when calculated from repeated random samples. Alternatively a confidence interval is provided for the parameter with a stated confidence coefficient. It is claimed that the unknown value of the parameter is contained in this interval. The end points of the interval depend upon the random sample, and the confidence coefficient is the proportion of times such confidence intervals will actually contain the unknown parameter in long sequence of repeated experiments. This coefficient is chosen to be of the order of 0.95 or 0.99. Statistical inference is not complete without some measure of precision of the procedure, e.g., the standard error or the confidence coefficient. In order to obtain these it is necessary that the sampling distribution of the statistic on which the procedure is based should be known under the specified model. It is so known when the model is parametric but is very rarely known under nonparametric models because of the fact that many families of c.d.f.'s are allowed under such models. With the help of a transformation it is possible to obtain estimates and confidence intervals for a certain kind of parameters in nonparametric models, but a theory of optimal estimation is possible under parametric models only. The theory of estimation of parameters which we develop in non-parametric inference will have the necessary accompaniment of the measures of precision, without the exact optimality properties like shortest confidence intervals, smallest variance, etc. which are available within parametric models.

A basic property that we like point estimation procedure to possess is consistency. If a statistic T_n (a function of observations $(X_1, X_2, \cdots, X_n)$) is an estimator of θ based on n observations then T_n is said to be consistent for θ if

$$\lim_{n \to \infty} P_\theta[|T_n - \theta| > \epsilon] = 0 \quad \forall \ \epsilon > 0.$$

This essentially says that consistent estimator T_n of θ must lie in a small interval around the true value of θ with probability approaching 1 as the number of observations (sample size) becomes large.

The statistics which have a completely known sampling distribution, even when the family of c.d.f.'s from which the observations arise is not known beyond the fact that it is a family of continuous c.d.f.'s, are known as distribution free statistics. Inference in nonparametric models has to be preferably based on such statistics. Lacking such statistics we look for those statistics whose sampling distribution is free of the family of c.d.f.'s from which the observations arise asymptotically, i.e., in the limit as the sample size $n \to \infty$. Such statistics are called asymptotically distribution free. The known asymptotic sampling distribution may be used in place of the exact sampling distribution, which is unknown, as an approximation in samples of not a very small size.

Sometimes the Law of Large numbers and the Central limit Theorem also show that certain common estimators are asymptotically distribution free. For example to estimate the mean of a distribution we often use $\bar{X}$, the sample mean. The classical central limit theorem tells us that if F has finite variance σ^2 then,

$$\sqrt{n} \frac{(\bar{X} - \mu)}{\sigma} \to N(0, 1) \text{ as } n \to \infty.$$

Then $100(1-\alpha)\%$ confidence interval for the population mean μ is given by

$$[\bar{X} - z_{1-\alpha/2} \frac{\sigma}{n}, \ \bar{X} + z_{\frac{1-\alpha}{2}} \frac{\sigma}{n}],$$

where $z_{1-\alpha/2}$ is the upper $100\frac{\alpha}{2}\%$ critical value of the standard normal distribution. If σ is not known, it may be replaced in the above by its estimate $s = \sqrt{\sum_{i=1}^{n}(x_i - \bar{x})^2/(n-1)}$, and the asymptotic properties of the confidence interval will hold by the combination of the Central limit Theorem and Slutsky's Theorem.

However, optimal estimation of parameters is natural only within finite dimensional parametric families and will not be pursued in this book. The concepts of sufficiency and completeness of statistics are also very important in various ways.

A statistic T_n (a function of observations $X_1, X_2, \ldots, X_n$) is said to be sufficient for θ if the conditional distribution of $X_1, X_2, \ldots, X_n$ given the value of T_n does not depend upon θ. Thus it is felt that once we have calculated T_n from the data, there is no additional information left in the sample which is relevant for inferences regarding θ.

A statistic T_n (or the family of sampling distribution of T_n) is said to be complete for θ if

$$E_\theta[g(T_n)] = 0 \ \forall \ \theta$$

implies that

$$g(T_n) \equiv 0, \quad \text{a.e.}$$

that is to say, if a function g of T_n has expectation $0 \ \forall \ \theta$ then the function must be uniformly 0.

The other major procedure of statistical inference is testing of hypotheses. Statistical hypotheses are statements regarding the statistical laws governing outcomes of random experiments (in terms of the distributions of the random variables). For example, it may be claimed that

$$H : X \sim N(\mu, \sigma^2), \quad -\infty < \mu < \infty, \; \sigma^2 > 0.$$

This is read as "the hypothesis is that the random variable X has a Normal distribution, with unspecified values of the mean μ and variance σ^2." Another statement could be

$H_1 : X$ has a symmetric distribution about μ, $\quad -\infty < \mu < \infty$

or

$H_2 : X \sim F$, $Y \sim G$ and $G(x) = F(x - \Delta)$, $\Delta > 0$.

The last statement says that "X and Y have the same probability distribution" except that values of X are shifted to the right by a positive constant Δ as compared to the values of Y.

There is very little difference between a model and a hypothesis, except that a model is accepted as true from prior considerations and the validity of the hypothesis is sought to be established with the help of observations. The common position of the statistician is that the experimenter can usually summarize the present knowledge regarding the statistical law as a hypothesis: it is called the null hypothesis. The experimenter has an alternative hypothesis in mind. It does not yet have the accepted status, but is only a tentative suggestion. These two hypotheses are set up as the null and alternative hypotheses:

H_0 : the currently accepted status

H_A : the tentative suggestion, perhaps based on some intuition, previous data, etc.

The purpose of a test of hypothesis is primarily to see whether the data suggests the rejection of the null hypothesis (current status) in favour of the alternative hypothesis (newly suggested by the experimenter). Any experimenter recognizes that the current status and the suggested alternative are not to be regarded as on par. The current status, being the established theory at this point of time, is to be viewed with some respect and to be abandoned for the new suggestion only if overwhelming experimental evidence is in favour of the new suggestion. The philosophy is that it is better to stick to the current status even if it is wrong, than to accept the new suggestion if it is possibly wrong. The burden of proof that the suggested alternative is the more appropriate statistical law is put on the experimenter. To make the ideas clear consider the following decision chart

Table 1.1

	H_0 is retained	H_0 is rejected in favour of H_A
H_0 is true	3	1
H_A is true	2	4

The statistical procedure has to take only one of the two actions (i) Reject H_0, (ii) Do not reject H_0.

Obviously the decisions represented by cells numbered 1 and 2 correspond to errors in decision making and those represented by cells numbered 3 and 4 refer to correct decisions. The decision 1, i.e., to reject H_0 when it is true is called the Type I error and the decision 2, i.e. to retain H_0 when the alternative is true is called Type II error. The error of Type I is to be regarded as much more serious than the error of Type II.

Since the actions are to be taken on the basis of data collected as outcomes of random experiments which are subject to variation from trial to trial, it is recognized that it is impossible to suggest rules which will completely eliminate errors. A statistical test of hypothesis is a rule which will have known probabilities of these errors. In fact a test would be so constructed that it will have specified probabilities of the errors of the two types. The concept of precision of a statistical inference procedure is crystalized through the probabilities of Type 1 and Type II errors.

In accordance with the asymmetry of the null and alternative hypothesis, the principle of statistical inference in this context is: keep the probability of Type 1 error at a low level (say, .01 or .05, etc.) and among all tests having this low level of probability of Type 1 error (called the significance level of the test) choose the test which minimizes in some way the probability of Type II error. This would ensure that the null hypothesis (which is the current status) will be wrongfully rejected only in a known low proportion of cases in long sequences of repetitions while doing our best to keep low the probability of Type II error.

Again the above prescription can be totally achieved only in relatively simple problems. Particularly the problems of such test construction have been solved with parametric models with only one unknown (real valued) parameter say θ. In case the null hypothesis is simple, i.e., specifying only a single probability distribution, say,

$$H_0 : \theta = \theta_0$$

and the alternative hypothesis is one sided, i.e.,

$$H_A : \theta > \theta_0$$

or

$$H'_A : \theta < \theta_0$$

then tests with minimum probability of Type II error for a specified probability of Type I error can be constructed in certain types of parametric families. In situations with more complex null and alternative hypotheses the construction of such optimal tests is not feasible. Then we, at all costs, maintain the fixed low level of significance, say α (i.e. probability of Type 1 error) and seek some other desirable properties which the test should possess when the alternative hypothesis is true. One such property is consistency of the test. It says that the probability of Type II error should tend to 0 as the number of observations tends to infinity. Or equivalently the power of the test (which is precisely probability of the Type II error subtracted from 1) should tend to 1 for any fixed distribution included in the alternative hypothesis as the sample size increases. This assures that while maintaining the probability of rejecting wrongfully the current status at a low acceptable level, the probability of accepting the alternative if in fact it is correct, tends to 1 with increasing sample size.

The use of fixed levels of significance has been criticized on the ground that there is no sanctity attached to these numbers such as 0.05 or 0.01. Even if the value of the statistic just falls short of the level .05 critical point we will not reject H_0. However, the same data would have led to its rejection if we had decided to use, say .055 as the level of significance. There may not have been any particular reason for choosing 0.05 over .055 as the level in the first place. Same comments apply to values of the statistic just exceeding the critical point. Hence, often the statistician reports the smallest level of significance at which the observed value of the statistic would have led to the rejection of the H_0. This information is adequate for making the rejection/acceptance decision at any fixed level of significance. Besides, it allows some further insight in the situation at hand through the largeness or smallness of this number. This number (the smallest level of significance at which the H_0 would be rejected) is called the P-value of the data.

In order to be able to fix the probability of Type I error, which is a probability to be calculated under the null hypothesis, we must know the sampling distribution of the test statistic under it. If the null hypothesis itself is a nonparametric family consisting of many parametric families of distribution functions, then usually it is not possible to have a single probability distribution for the statistic which is to be used as the test statistic. In some situations the probability integral transformation discussed in Chapter 2 saves the day. After transforming the data through it we see that the probability of certain subsets of the sample space called rank sets can be calculated under the null hypothesis and tests based on these rank tests will have the specified levels of significance. Such a test which retains its level of significance for a nonparametric null hypothesis, i.e., one which encompasses many parametric families of distribution functions, is called a distribution free test. However, it is practically impossible to obtain the exact probability of Type II error or the power of any such test; leave alone choosing the test with maximum power among all such tests for all alternatives included in the alternative. Then tests which are intuitively felt to be able to discriminate between the null and the alternative hypotheses are considered. Also certain weaker optimality criteria like locally most powerful rank (LMPR) structure are brought to bear on the problem. Help is taken of the invariance principle, if relevant. In the subsequent chapters these concepts will be properly introduced and used for development of tests in nonparametric models.

The above discussion clearly indicates that the problem of testing of hypotheses in the context of nonparametric models is too complex to lead to unique optimal tests and in any given situation there may exist several useful tests without any clear preference for the entire alternative hypothesis. Then measures of comparison have to be brought into play to make a choice. The concept of asymptotic relative efficiency (ARE) as introduced by Pitman and further developed by Noether is a very useful tool. It provides a single number which summarizes the comparison of two tests of the same null hypothesis when used against certain parametric sequences of alternatives which tend to the null hypothesis. We shall develop the concept of ARE and its use for comparison of tests in the Appendix.

1.6   Maximum Likelihood Estimation

A principle of estimation which has been proved to be extremely valuable in parametric families is the principle of maximum likelihood estimation. Generalizations of this principle have been found useful in nonparametric families while estimating entire distribution functions. Hence we describe here this principle in parametric situations briefly and then explain its generalizations.

Let θ be the identifying parameter for the class of p.d.f.'s $\{f_\theta, \theta \in \Theta\}$, Θ is the parametric space which is a subset of R. The data is obtained through a random experiment whose outcomes are governed by the p.d.f. $f_\theta(x)$, θ being unknown and required to be estimated. The data consists of the random sample $x_1, x_2, \cdots, x_n$, the realization of n independent random variables $X_1, X_2, \cdots, X_n$ with identical density $f_\theta(x)$. The joint p.d.f. of these rv.'s is $\prod_{i=1}^n f_\theta(x_i)$ and one may regard $\prod_{i=1}^n f_\theta(x_i)dx_i$ as the probability element of the observed data. Now we don't know θ; we have to estimate it. It is some value in Θ. We choose to accept that value of θ which provides the maximum value of $\prod_{i=1}^n f_\theta(x_i)$ for the given data $x_1, x_2, \cdots, x_n$. In other words, loosely speaking, we estimate θ by $\hat{\theta}$ if we feel that the p.d.f. $f_{\hat{\theta}}(x)$ gives the larger probability element to the observed data among all possible models. The principle of maximum likelihood essentially states that choose $\hat{\theta}$ to be the estimator of θ if

$$L(\hat{\theta}|x_1, \cdots, x_n) = \prod_{i=1}^n f_{\hat{\theta}}(x_i)$$

$$= \sup_\theta L(\theta|x_1, \cdots, x_n)$$

$$= \sup_\theta \prod_{i=1}^n f_\theta(x_i).$$

The function

$$L(\theta|x_1, \cdots, x_n) = \prod_{i=1}^n f_\theta(x_i)$$

as a function for the given data $x_1, \cdots, x_n$ is called the likelihood function of θ and $\hat{\theta}$ the maximum likelihood estimator (MLE) of θ. One may paraphrase the principle of maximum likelihood by saying that given several models for the outcomes of the experiment, we choose the one under which the observed data has the maximum likelihood of appearing. We cannot bring ourselves to accept as the correct model one which gives less probability to the observed data than some other model. In the very simple experiment, say tossing a coin, suppose that we observe a head on a single toss. Now if the choice of probabilities of a head is between $\frac{1}{3}$ and $\frac{1}{2}$ ($\theta = \frac{1}{3}$ or $\frac{1}{2}$) then we will go for $\hat{\theta} = \frac{1}{2}$ as this gives larger probability to the observed data (a head).

Such estimators exist and are consistent with limiting Normal distribution under very weak regularity conditions, but in some pathological cases they may not exist or even if they do exist, they are seen to have undesirable properties.

There are other principles of estimation such as minimum variance unbiased estimation, etc., which will not be detailed here.

1.7   Observation with Censored Data

So far we have considered only one scheme of data collection, namely a
random sample or a collection of the value of independent and identically
distributed random variables with distribution F. In many experiments involving lifetimes, such as reliability studies of engineering/mechanical systems or survival studies in biomedical situations, schemes which entail censored observations are used or are inherent in the situation. Nonparametric
and semiparametric methods have been successfully adapted and developed
for such data.

In studies involving lifetimes, the time taken for experimentation can be
too long and is always uncertain. Type I and Type II censoring is adopted
at the design stage to have (a) a fixed time and (b) a comparatively shorter
time, respectively for experimentation.

1.7.1   *Type I (Time) Censoring*

Suppose n statistically independent and identical items are put into operation (or put on test) simultaneously for investigating the common statistical
law governing their lifetimes. The classic data collection scheme would be
to continue until all the items have failed thus yielding the random sample
of size n from F. But to have a fixed (and often shorter) time for experimentation, the scheme is to terminate the experiment after a fixed time t_0,
only record the lifetimes which were shorter than or equal to t_0 and record
for the remaining items the fact that their lifetimes were greater than t_0,
or in other words, to say that they were censored at time t_0. Let f be
the p.d.f. corresponding to the c.d.f. F. We denote by t_i the lifetime of
the i-th item, if it is observed; $\delta_i = 1$ if the lifetime is observed and 0 if it
is censored; and of course t_0, the censoring time. The likelihood of the n
observations thus made is

$$L(f, t_0 | t_1, \cdots, t_n, \quad \delta_1, \delta_2, \cdots, \delta_n)$$
$$= \prod_{i=1}^{n} \{f(t_i)\}^{\delta_i} \{1 - F(t_0)\}^{1-\delta_i},$$
$$0 < t_i \leq t_0.$$

1.7.2   *Type II (Order) Censoring*

In this scheme also n items are put on test and the experiment is termi-
nated as soon as k $(1 \leq k \leq n)$ items have failed. The integer k is chosen in

advance of the experimentation. Thus the time for which the experiment continues is again random, but shorter than what will be needed for observing the failure of all the n items. This scheme is adopted to eliminate the possibility, which exists in Type I scheme, of not recording any completed lifetime at all. Since the data is in the form of the k smallest order statistics of a random sample of size n, the likelihood may be written as the joint density of these

$$L(f|x_{(1)}, x_{(2)}, \cdots, x_{(k)})$$

$$= \frac{n!}{(n-k)!} \prod_{i=1}^{k} f(x_{(i)}) \{1 - F(x_{(k)})\}^{n-k},$$

$$0 \le x_{(1)} \le \cdots \le x_{(k)} < \infty.$$

Since we are dealing with life times, which are positive valued random variables, we take the lower limit as 0 rather than $-\infty$ in both cases discussed above and also in the case mentioned below.

1.7.3 *Type III (Random Censoring)*

This type of censoring is more in the nature of experiments than a matter of design. It is observed that in lifetime studies on human or other biological entities it is comparatively difficult to get the requisite number of items for experimentation at the same time. Also, such subjects after entering the study, may be withdrawn before the completion of the lifetime (i.e. death) from the study. These withdrawals will not necessarily occur at the same age for every subject. If a study of the lifetime of patients under terminal cancer is organized, it may be that some patients are withdrawn from the study (for whatever reasons, e.g., trying alternative therapy), die of some cause other than cancer (competing risks), etc.

The situation is modelled by two random variables: a lifetime X and a censoring time C. For each subject one of these yields a realization, X, if the lifetime is observed and C, if it is censored prior to the completion. Thus the observations on n such independent subjects are (T_i, δ_i), $i = 1, 2, \cdots, n$ where

$$T_i = \min(X_i, C_i),$$

and

$$\delta_i = I[X_i \le C_i] = \begin{cases} 1 & \text{if } X_i \le C_i \\ 0 & \text{if } X_i > C_i, \end{cases}$$

are the actual recorded observation and the indicator of whether it was uncensored or a censored life. The simplest model then assumes that X and C are independent with distributions F and G (densities f and g) leading to the following likelihood function

$$L(f, g | t_1, t_2, \ldots, t_n) = \prod_{i=1}^{n} \{f(t_i)\overline{G}(t_i)\}^{\delta_i} \{g(t_i)\overline{F}(t_i)\}^{1-\delta_i}$$

where $\overline{F}(x) = 1 - F(x)$ and $\overline{G}(x) = 1 - G(x)$, respectively.

If we take the censoring time distribution $G(x)$ to be degenerate at t_0, a fixed time of censoring, then Type III censoring reduces to Type I censoring. Under the conditions stated above Type III censoring does not reduce to Type II censoring since the censoring time in Type II, which is $X_{(k)}$ the k-th order statistic of $X_1, X_2, \cdots, X_n$ the n lifetimes, is not independent of them.

1.8   Ranked Set Sampling

The Ranked Set Sampling scheme was introduced by [McIntyre (1952)]. For a while it did not attract the attention it deserved, but there has been a surge in the interest in it over the last twenty years or so. [Patil *et al.* (1994)], [Barnett (1999)], [Chen *et al.* (2003)], [Wolfe (2004)] may be seen as some of the landmark contributors.

Let us first describe the basic framework of this methodology. It consists of the following stages. Suppose we are interested in a characteristic represented by the values of a random variable X taken by each of the units in the population. Let X have a continuous probability distribution with c.d.f. F and p.d.f. f. Let μ be the expectation of X and σ^2 its variance.

First choose n units from the population as a simple random sample. By some (hopefully inexpensive) procedure select the unit with the smallest value of X. Let it be $X_{[1]}$.

A second independent simple random sample of size n is chosen and the unit with the second smallest value $X_{[2]}$ of the character is selected. Similarly, continue this process until one has $X_{[1]}, X_{[2]}, \ldots, X_{[n]}$, a collection of independent order statistics from n disjoint collections of n simple random samples.

This constitutes the basic Ranked Set Sample. These are independently distributed with respective p.d.f.s

$$f_{(i)}(x_{[i]}) = \frac{n!}{(i-1)!(n-i)!} F(x_{[i]})^{i-1}(1-F(x_{[i]})^{n-i}f(x_{[i]}), \quad -\infty < x_{[i]} < \infty.$$

$$(1.8.1)$$

Note that we obtained n^2 simple random samples from the population, but retained only one (with the appropriate rank) from each group of n samples. So the total effort has been to rank n random samples of size n each, and retaining only the i-th $(i = 1, 2, \ldots, n)$ order statistic from the i-th group respectively, which are to be actually measured.

It is expected that ranking is cheap and measuring is expensive. Ranking n^2 observations and using n of them after measurement is thus a sampling method of boosting the efficiency of the statistical procedures from that based merely in groups of n simple random sample measurements.

In order to retain the reliability of the rankings, it is usually suggested that it be carried out in small groups, say upto $4, 5, 6$ in size. However, this would limit both the versatility and efficiency of the procedures. Hence recourse is taken to replicating the whole process m times. Thus mn^2 units are examined, each group of n is ranked within itself and the i-th order statistic is obtained from m groups, leading to the data

$$X_{[i]j}, i = 1, 2, \ldots, n \; j = 1, 2, \ldots, m.$$

This constituted balanced RSS sampling. In unbalanced sampling instead of m replications for each order statistic, one will have m_i replications of the i-th order statistic.

Parametric estimation and testing of hypothesis is well studied under this scheme of sampling. See, for example, [Chen *et al.* (2003)]. Some nonparametric procedures and their properties too have been discussed in the literature. We shall include some of these in the appropriate chapters of the book.

1.8.1 *Estimation of the Mean*

Let us now consider the estimation of μ, the mean of F. If we use k SRS observations then $\bar{X}$, its sample mean, is an unbiased estimator with variance σ^2/k. The mean

$$\bar{X}^* = \frac{1}{n}\sum_{i=1}^{n} X_{[i]}$$

of the k RSS observations is also unbiased for μ as seen below.

$$E(\bar{X}^*) = \frac{1}{n} \sum_{i=1}^{n} E(X_{[i]})$$

$$= \frac{1}{n} \sum_{i=1}^{n} \int_{-\infty}^{\infty} x \frac{n!}{(i-1)!(n-i)!} F(x)^{i-1} (1 - F(x))^{n-i} f(x) dx$$

$$= \int_{-\infty}^{\infty} x \left[\sum_{i=1}^{n} \binom{n-1}{i-1} F(x)^{i-1} (1 - F(x))^{n-i} \right] f(x) dx$$

$$= \int_{-\infty}^{\infty} x f(x) dx$$

$$= \mu.$$

1.9   Introduction to Bayesian Inference

It is given in statistical inference that different statisticians working with the same procedure with well defined properties will come to the same conclusion with the same data. This is called 'objective inference' and is a characteristic of the so called 'frequentist' approach. On the other hand, the Bayesian approach allows inputs from the statistician (or the scientist at the back of the statistician) in the inference procedure. This is said to be 'subjective inference' or 'Bayesian inference' since the basic setup in the procedure depends on the use of the famous Bayes theorem given below.

Let there be two events A and B with positive probabilities. Then

$$P(A|B)P(B) = P(A \cap B) = P(B|A)P(A),$$

which leads to

$$P(B|A) = \frac{P(A|B)P(B)}{P(A)}.$$

Now consider the problem of estimation of a parameter θ on the basis of data $\underline{x} = (x_1, x_2, \ldots, x_n)$ coming from a probability density $f(\underline{x}|\theta)$. The likelihood principle described earlier says that regard $f(\underline{x}|\theta) = L(\theta|\underline{x})$ as a function of θ for the given data and estimate θ by $\hat{\theta}$ which is the value of θ for which $L(\theta|\underline{x})$ is maximized. In this procedure there is no scope for the statistician to incorporate his personal (i.e. subjective) beliefs. But in many cases the statistician/scientist does have preferences about the true value of the parameter θ. The Bayesian paradigm proposes that the preferences or degrees of belief be summarized by a probability distribution $p(\theta)$: the greater the value of $p(\theta)$, the greater is the degree of belief in the corresponding value of θ. So in a manner of speaking θ is regarded as a random variable, and the density $f(\underline{x}|\theta)$ a conditional density of the data given the value of the random variable θ. Then by a simple application of the extended version of Bayes theorem we can write

$$f(\underline{x}|\theta)p(\theta) = f(\underline{x}, \theta) = p(\theta|\underline{x})m(\underline{x})$$

and

$$p(\theta|\underline{x}) = \frac{f(\underline{x}|\theta)p(\theta)}{m(\underline{x})}.$$

Here $m(\underline{x})$ is the marginal p.d.f. of $\underline{x}$ given by $\int_\Theta f(\underline{x}|\theta)p(\theta)d\theta$, where Θ is the parametric space. The two p.d.f.'s $p(\theta)$ and $p(\theta|\underline{x})$ are called the prior and posterior density of θ, that is the degrees of belief the statistician holds initially and the modified degrees after the observation of the data $\underline{x}$. A point estimate of θ can be taken to be the mode of $p(\theta|\underline{x})$ that is the value where the posterior distribution is maximum (rather than where $L(\theta)$ is maximum). The name Bayes estimator is given to the expectation $E(\theta|\underline{x})$ of the posterior distribution. The first of the estimators, the mode, retains the maximum degree of (posterior) belief, where as the Bayes estimator is the one which minimizes the posterior risk, i.e., the expected squared posterior error

$$\int_\Theta (\theta - \tilde{\theta})^2 p(\theta|\underline{x})d\theta,$$

which is minimized at $\tilde{\theta} = E(\theta|\underline{x})$ among all values of $\tilde{\theta}$ in Θ.

Similar analysis can be done with respect to other loss functions also. The above is a parametric treatment. In a later chapter we see that use of the Bayesian principle in a nonparametric set up is much more difficult and computer intensive.

1.10 Remarks

Other statistical problems of interest include ranking, selection, and the vast array of multivariate problems. Again the development of nonparametric methods for these problems has been rather limited as we lack adequate information for calculating exact error rates which indicate the precision of the procedures.

By now the above discussion brings out the fact that estimation procedures and tests of hypothesis, under nonparametric models, should be based on distribution free statistics. If appropriate distribution free statistics are not forthcoming then we look for asymptotically distribution free statistics so that the precision of the estimators and the level of significance of the tests may at least approximately be known.

Even in case of most distribution free statistics the exact sampling distribution, though known, become very complicated with increasing sample size. We then obtain the limiting sampling distributions of the statistics for use as approximations to the exact distribution.

Details of principles of statistical inference, methods of estimation and testing hypothesis may be found in [Lehmann (1991)], [Lehmann and Romano (2006)], [Fraser (1957)], [Rohatgi and Saleh (2015)], [Wilks (1962)] etc.

We should mention some well-known books on Nonparametric methods. The earlier rather theoretical books are [Fraser (1957)], [Hájek and Šidák (1967)], [Puri *et al.* (1971)] and [Randles and Wolfe (1979)]. The later books catering for applied statisticians are [Conover and Conover (1980)] and ([Sprent and Smeeton (2016)], fourth edition). Some more recent books [Govindarajulu (2007)], [Wasserman (2007)], [Gibbons and Chakraborti (2011)] and [Hollander *et al.* (2013)] contain a blend both theory and methods.

Chapter 2

ORDER STATISTICS

2.1 Random Sample from a Continuous Distribution

If a random experiment is performed n times in such a way that the outcomes in any trial do not affect those in other trials then we say that we have n statistically independent trials. If $X_1, X_2, \cdots, X_n$ are the random variables representing the outcomes then these would have a common continuous c.d.f. $F(x)$ with p.d.f. $f(x)$. Suppose F is unknown; only known to belong to the class $\mathcal{J}$ of all continuous distributions. This is a nonparametric model. Symbolically we write $X_1, X_2, \cdots, X_n \sim$ i.i.d. random variables from F. The actual realizations of these random variables after the experiments have been performed will be denoted by lower case letters $x_1, x_2, \cdots, x_n$. These are real numbers being the observed values of the random variables $X_1, X_2, \cdots, X_n$ respectively, or the realized random sample.

2.2 Order Statistics

We define new random variables $X_{(1)}, X_{(2)}, \cdots, X_{(n)}$ in the following way:

$$X_{(1)} = \min(X_1, X_2, \cdots, X_n),$$

i.e., the smallest among all observations,

$$X_{(2)} = \min(\{X_1, X_2, \cdots, X_n\} - X_{(1)}),$$

i.e., the smallest among all observations except $X_{(1)}$,

$$X_{(i)} = \min(\{X_1, X_2, \cdots, X_n\} - \{X_{(1)}, X_{(2)}, \cdots, X_{(i-1)}\}), \quad i = 2, \ldots, n-1,$$

$$X_{(n)} = \max(X_1, X_2, \cdots, X_n).$$

$(X_{(1)}, X_{(2)}, \cdots, X_{(n)})$ are called order statistics of the random sample $X_1, X_2, \cdots, X_n$.

Their observed values are the ordered values $x_{(1)} \leq x_{(2)} \leq \cdots \leq x_{(n)}$ of the realized observations $x_1, x_2, \cdots, x_n$. These are called the sample order statistics. We also call $X_{(i)}$ and $x_{(i)}$, $i = 1, 2, \ldots, n$, the i-th order statistic and the i-th sample order statistic, respectively.

Once the order statistics are computed, the information as to which was the observation number one, number two, etc., is lost. So it is a reduction of the data in this sense. One may also notice that the vector of observations $(x_1, x_2, \cdots, x_n)$ can denote any point in the space R^n, provided the original observation $x_i's$ are in R. But the vector of order statistics

$$(x_{(1)}, x_{(2)}, \cdots, x_{(n)}) \equiv (y_1, y_2, \cdots, y_n)$$

can denote any point only in the subset of R^n

$$\{ \underline{y} \in R^n \quad \text{and} \quad y_1 \leq y_2 \leq \cdots \leq y_n \}$$

which is much smaller than R^n itself.

2.3   Sampling Distribution of Order Statistics

The joint density of the observation $X_1, X_2, \cdots, X_n$ is

$$f(x_1, x_2, \cdots, x_n) = \prod_{i=1}^{n} f(x_i), \quad -\infty < x_1, x_2, \ldots, x_n < \infty. \qquad (2.3.1)$$

As already explained the range of the order statistics $X_{(1)}, X_{(2)}, \cdots, X_{(n)}$ is $-\infty < x_{(1)} \leq x_{(2)} \leq \cdots \leq x_{(n)} < \infty$. Let A^1 be an arbitrary measurable set in this range and $g(x_{(1)}, x_{(2)}, \cdots, x_{(n)})$ be the joint density of the order statistics. Then,

$$P[(X_{(1)}, \cdots X_{(n)}) \in A^1] = \int \cdots \int_{A^1} g(x_{(1)}, \cdots, x_{(n)}) dx_{(1)} \cdots dx_{(n)}. \qquad (2.3.2)$$

Let $A^1, A^2, \cdots, A^{n!}$ be the sets in R^n obtained by permuting the coordinates of points in A^1. Observe that the same value of order statistics is generated by all the permutations of $(x_1, x_2, \cdots, x_n)$ and the joint density function is symmetric in its arguments. We can write

$$P[(X_{(1)}, \cdots, X_{(n)}) \in A^1] = \sum_{j=1}^{n!} P[(X_1, X_2, \cdots, X_n) \in A^j]$$

$$= \sum_{j=1}^{n!} \int \cdots \int_{A^j} \prod_{i=1}^{n} f(x_i) dx_i$$

$$= n! \int \cdots \int_{A^1} \prod_{i=1}^{n} f(x_i) dx_i. \qquad (2.3.3)$$

Comparison of (2.3.2) and (2.3.3) tells us that the joint density of the order statistics $(X_{(1)}, X_{(2)}, \cdots, X_{(n)})$ is

$$g(x_{(1)}, \cdots, x_{(n)}) = n! \prod_{i=1}^{n} f(x_{(i)}), \quad -\infty < x_{(1)} \leq \cdots \leq x_{(n)} < \infty. \quad (2.3.4)$$

This expression for the joint density easily gives, by direct integration, the marginal density of one, two or more order statistics. We obtain the marginal density of $X_{(i)}$ as

$$g_{X_{(i)}}(x) = \frac{n!}{(i-1)!(n-i)!} f(x)(F(x))^{i-1}(1 - F(x))^{n-i}, -\infty < x < \infty. \tag{2.3.5}$$

The joint density of $X_{(i)}$ and $X_{(j)}$ for $i < j$ is

$$g_{X_{(i)}, X_{(j)}}(x, y) = [\frac{n!}{(i-1)!(j-i-1)!(n-j)!}]$$
$$[f(x)f(y)(F(x))^{i-1}(F(y) - F(x))^{j-i-1}(1 - F(y))^{n-j}],$$
$$-\infty < x < y < \infty. \tag{2.3.6}$$

In particular the marginal density of the smallest order statistics $X_{(1)}$ is

$$g_{X_{(1)}}(x) = nf(x)(1 - F(x))^{n-1}, \quad -\infty < x < \infty, \tag{2.3.7}$$

and that of the largest order statistics $X_{(n)}$ is

$$g_{X_n}(x) = nf(x)[F(x)]^{n-1}, \quad -\infty < x < \infty. \tag{2.3.8}$$

The joint density of $X_{(1)}$ and $X_{(n)}$ is

$$g_{X_{(1)}, X_{(n)}}(x, y) = n(n-1)f(x)f(y)[F(y) - F(x)]^{n-2},$$
$$-\infty < x < y < \infty. \tag{2.3.9}$$

The c.d.f. of $X_{(i)}$ is seen to be

$$
\begin{aligned}
G_{X_{(i)}}(x) &= P[X_{(i)} \leq n] \\
&= P[\text{at least } i \text{ of } X_1, X_2, \cdots, X_n \\
&\quad \text{ are less than or equal to } x] \\
&= \sum_{j=i}^{n} P[\text{exactly } j \text{ of } X_1, X_2, \cdots, X_n \\
&\quad \text{ are less than or equal to } x] \\
&= \sum_{j=i}^{n} \binom{n}{j} (F(x))^j [1 - F(x)]^{n-j}, \quad -\infty < x < \infty.
\end{aligned}
$$

$$(2.3.10)$$

In particular,

$$
G_{X_{(1)}}(x) = 1 - (1 - F(x))^n,
$$

and

$$
G_{X_{(n)}}(x) = (F(x))^n. \tag{2.3.11}
$$

Differentiating $G_{X_i}(x)$ as obtained in (2.3.10) leads to the same expression for $g_{X_{(i)}}(x)$ as obtained in (2.3.5).

2.4 Sufficiency and Completeness of Order Statistics

The concept of sufficiency provides a useful summarization of the data without losing any statistically relevant information. There can be many sufficient statistics for the same statistical family. For the family of all absolutely continuous distributions, or for any subfamily of it, the order statistics are a sufficient statistics. This is proved by observing that,

$$
P[(X_1, X_2, \cdots, X_n) \in A | X_{(1)} = x_{(1)}, \cdots, X_{(n)} = x_{(n)}] = \frac{m(A)}{n!},
$$

where $m(A)$ is the number of points, from the $n!$ points whose coordinates are given by the permutations of $(x_{(1)}, x_{(2)}, \cdots, x_{(n)})$, included in A. A complete proof may be seen in [Fraser (1957)]. As this conditional probability does not depend upon the probability distribution of $X_1, X_2, \cdots, X_n$, we have sufficiency for $X_{(1)}, X_{(2)}, \cdots, X_{(n)}$.

Note that order statistics are sufficient statistics of dimension n which increases with the sample size.

Suppose $\mathcal{F}$ is the class of all absolutely continuous distributions. It can also be seen from [Fraser (1957)] that if

$$E[g(X_{(1)}, \cdots, X_{(n)})] = 0 \ \forall \ f \in \mathcal{F},$$

then

$$g(X_{(1)}, \cdots, X_{(n)}) \equiv 0. \tag{2.4.12}$$

Hence order statistics are also complete for the family of all absolutely continuous distribution functions.

In statistical theory it is well know that if we can have a statistic which is a function of a complete and sufficient statistic then it is the unique minimum variance unbiased estimator (UMVUE) of its expectation. See Rao - Blackwell and Lehmann - Scheffé theorems, say, in [Lehmann (1991)], [Casella and Berger (2002)]

2.5 Probability Integral Transformation

Suppose the random variable X has a continuous c.d.f. $F(x)$. Consider the transformation

$$Y = F(X). \tag{2.5.13}$$

Then Y is a random variable since $F : R \to (0, 1]$ is a monotone increasing (also right continuous) transformation. It is easy to see that if F is a continuous c.d.f. then Y itself has the uniform distribution over $(0, 1]$. Let $G(y)$ be the c.d.f. of Y. Then, for $0 \leq y < 1$

$$\begin{aligned} G(y) &= P[Y \leq y] \\ &= P(X \leq F^{-1}(y)] \\ &= FF^{-1}(y) = y. \end{aligned}$$

Hence,

$$G(y) = \begin{cases} 0 & \text{if } y < 0, \\ y & 0 \leq y < 1, \\ 1 & \text{if } y \geq 1. \end{cases}$$

Here

$$F^{-1}(y) = \inf\{x : F(x) \geq y\}.$$

This transformation is useful in specifying the probability of X being in certain sets even if we do not know its distribution function F, as long as it is continuous. For example, consider two i.i.d. random variables X_1 and X_2 both with the same distribution function F. Then consider the probability of the set $\{(x_1, x_2) : -\infty < x_1 \leq x_2 < \infty\}$ in R^2 or $P[X_1 \leq X_2]$. This set is invariant under any continuous monotone increasing transformation $(g(x_1), g(x_2))$ applied to (x_1, x_2), i.e.

$$\{(x_1, x_2) : -\infty < g(x_1) \leq g(x_2) < \infty\} \equiv \{(x_1, x_2) : -\infty < x_1 \leq x_2 < \infty\}.$$

Hence,

$$P[g(X_1) \leq g(X_2)] = P[X_1 \leq X_2].$$

If we specify $g = F$ then

$$P[X_1 \leq X_2] = P[F(X_1) \leq F(X_2)]$$
$$= P[Y_1 \leq Y_2],$$

where Y_1 and Y_2 are independent random variables with $U(0, 1)$ distributions. Hence,

$$P[X_1 \leq X_2] = P[Y_1 \leq Y_2] = \int_0^1 \int_0^{x_2} dx_1 dx_2 = \frac{1}{2}.$$

Another useful transformation is given by

$$Z = -\log[1 - F(X)]. \tag{2.5.14}$$

Since $Y = F(X)$ has $U(0, 1)$ distribution, $1 - F(X) = 1 - Y$ too has $U(0, 1)$ distribution.

$$P[Z \leq z] = P[-\log(1 - Y) \leq z] = P(1 - Y \geq e^{-z}) = 1 - e^{-z}, \ 0 \leq z < \infty.$$

That is to say that the transformed random variable Z has the exponential distribution with mean 1.

Comments:

(i) One can obtain the probability of sets which are invariant under continuous monotone transformations in the above manner by exploiting the properties of the probability integral transformation given in (2.5.13).

(ii) These results are of great value in deriving sampling distributions of certain statistics when we only know that the data has arisen from continuous probability distributions.

(iii) We deliberately construct statistics which depend upon X through the transformations (2.5.13) and (2.5.14) which are denoted above by Y and Z, respectively, so that the sampling distributions are tractable. And they are also distribution-free as they do not depend on the choice of underlying c.d.f. F, as long as it is continuous.

(iv) The probability integral transform is used to generate random variables from non uniform distributions. For example, the c.d.f. of exponential random variable with mean 1 is $F(x) = 1 - e^{-x}$. Then, $Y = 1 - e^{-X}$ is a $U(0, 1)$ random variable. Hence, $x = -log_e(1 - y)$ is an observation from standard exponential distribution. Or, in general, if y is an observation from $U(0, 1)$, then $x = F^{-1}(y)$ will be an observation from the c.d.f. F.

2.6 Order Statistics of Uniform and Exponential Random Samples

In section 2.5 we saw that certain simple transformations take an arbitrary continuous random variable to the uniform and exponential random variables. In this section we derive the expressions for the p.d.f. and other functions of order statistics from these random samples.

2.6.1 *Uniform Distribution*

Let $X_1, X_2, \ldots, X_n$ be a random sample from $U(0, 1)$ distribution. The density of the $U(0, 1)$ random variable is

$$f(x) = 1, \quad \text{if} \quad 0 < x < 1$$
$$= 0, \quad \text{elsewhere.} \qquad (2.6.15)$$

We get the joint density of order statistics from uniform distribution is

$$g(x_{(1)}, x_{(2)}, \cdots, x_{(n)}) = n!, \quad 0 < x_{(1)} \leq \cdots \leq x_{(n)} < 1, \qquad (2.6.16)$$

and the marginal density of the *ith* order statistics $X_{(i)}$ from $U(0, 1)$ is

$$g_{X_{(i)}}(x) = \frac{n!}{(i-1)!(n-i)!} x^{i-1}(1-x)^{n-i}, \quad 0 < x < 1, \qquad (2.6.17)$$

which is the Beta $(i, n - i + 1)$ distribution. Hence,

$$E(X_{(i)}) = \frac{i}{n+1},$$

and

$$Var(X_{(i)}) = \frac{i(n+1-i)}{(n+2)(n+1)^2}.$$

The joint density of the *ith* and the *jth* order statistics $X_{(i)}$ and $X_{(j)}$ for $i < j$ from $U(0,1)$ distribution is given by

$$g_{X_{(i)},X_{(j)}}(x,y) = \frac{x!}{(i-1)!(j-i-1)!(x-j)!} x^{i-1}(y-x)^{j-i-1}(1-y)^{x-j},$$

$$0 < x \le y < 1 \qquad (2.6.18)$$

and

$$Cov(X_{(i)}, X_{(j)}) = \frac{i(n+1-j)}{(n+1)^2(n+2)}. \qquad (2.6.19)$$

In particular,

$$Cov(X_{(1)}, X_{(n)}) = \frac{1}{(n+1)^2(n+2)},$$

$$Cor(X_{(1)}, X_{(n)}) = \frac{1}{n}.$$

Hence, correlation between $X_{(1)}$ and $X_{(n)}$ is inverse of the sample size n.

Substituting $i = 1$ and $j = n$ in (2.6.18) we obtain the joint distribution of the smallest and the largest order statistics of the uniform random sample, as

$$g_{X_{(1)},X_{(n)}}(x,y) = n(n-1)(y-x)^{n-2}, \quad 0 < x \le y < 1. \qquad (2.6.20)$$

Consider the transformation

$$R = X_{(n)} - X_{(1)}, \quad S = X_{(1)}.$$

Integrating the joint density of (R, S) with respect to S will give us the sampling distribution of R, the sample range, in the case of a random sample from $U(0,1)$ distribution as,

$$g_R(r) = n(n-1)r^{n-2}(1-r), \quad 0 < r < 1. \qquad (2.6.21)$$

Hence, the sample range has Beta$(n-1, 2)$ distribution.

2.6.2   *Exponential Distribution*

Let $X_1, X_2, \ldots, X_n$ be i.i.d. exp(1) random variables. Then the p.d.f. of X_1 is

$$f(x) = e^{-x}, \quad 0 < x < \infty, \qquad (2.6.22)$$

and the c.d.f. of X_1 is

$$F(x) = 1 - e^{-x}, \quad 0 < x < \infty. \qquad (2.6.23)$$

Then the joint density of n order statistics from $\exp(1)$ is

$$g_{X_{(1)},\cdots,X_{(n)}}(x_{(1)}, x_{(2)}, \cdots, x_{(n)})$$

$$= n! \prod_{i=1}^{n} e^{-x_{(i)}}, \quad 0 < x_{(1)} \leq \cdots \leq x_{(n)} < \infty, \tag{2.6.24}$$

and the pdf of the *ith* order statistic $X_{(i)}$ from $\exp(1)$ is

$$g_{X_{(i)}}(x) = \frac{n!}{(i-1)!(n-i)!}(1 - e^{-x})^{i-1}e^{-(n-i+1)x}, 0 < x < \infty. \tag{2.6.25}$$

Consider the following 1:1 transformation

$$Y_i = X_{(i)} - X_{(i-1)}, \quad i = 1, 2, \cdots, n,$$
$$X_{(0)} = 0.$$

Then, the joint p.d.f. of $Y_1, Y_2, \ldots, Y_n$ is

$$f_{y_1,\cdots,y_n}(y_2, \cdots, y_n) = n!e^{-\sum_{i=1}^{n}(n-i+1)y_i}, 0 < y_1, y_2, \ldots, y_n < \infty. \tag{2.6.26}$$

This shows that $Y_1, Y_2, \cdots, Y_n$ are independently and exponentially distributed with means $\frac{1}{n-i+1}, i = 1, 2, \cdots, n$.

Hence $Z_i = (n-i+1)Y_i, i = 1, \cdots, n$ which are called normalized sample spacings are once again independent and identically distributed $\exp(1)$ random variables. These are of particular interest in reliability theory, leading up to the concept of total time on test.

Also these results lead to

$$E(X_{(i)}) = E\left(\sum_{j=1}^{i}Y_j\right) = \sum_{j=1}^{i}\frac{1}{(n-j+1)}$$

and

$$Var(X_{(i)}) = Var\left(\sum_{j=1}^{i}Y_j\right) = \sum_{j=1}^{i}\frac{1}{(n-j+1)^2}.$$

When the random sample is not from the uniform or exponential distribution, it is usually not possible to obtain tractable expressions for the p.d.f. or even the moments of the order statistics.

Comments:

(i) Using the 'delta' method one can write down approximate expressions for the moments when the original distribution is an arbitrary continuous distribution with c.d.f. F and p.d.f. f. The first two moments approximately are:

$$E(X_{(i)}) \approx F^{-1}\left(\frac{i}{n+1}\right),$$

and

$$V(X_{(i)}) \approx \frac{i(n-i+1)}{(n+1)^2(n+2)}\left[f\left\{F^{-1}\left(\frac{i}{n+1}\right)\right\}\right]^{-1}.$$

(ii) Under regularity conditions the asymptotic distribution of the standardized version of $X_{(i)}$ as $n \to \infty$ in such a way that $\frac{i}{n} \to p$, $0 < p < 1$ is standard Normal, that is to say,

$$\left(\frac{p(1-p)}{nf^2(F^{-1}(p))}\right)^{-1/2}(X_{(i)} - F^{-1}(p)) \to N(0,1) \text{ as } n \to \infty.$$

In case $\frac{i}{n} \to 0$ or 1, as $n \to \infty$ then the above normality does not hold and the limiting distribution (with different normalization) is one of three possible extreme value distributions. See e.g. [Gumbel (1958)].

(iii) If we look at this asymptotic result carefully, the asymptotic variance of the i-th order statistic is of order $\frac{1}{n}$, that is, tending to 0 as $n \to \infty$. Hence, the i-th order statistic is a consistent estimator of its asymptotic expectation $F^{-1}(p)$, which is the quantile of order p of the distribution F.

(iv) If we were to base confidence intervals on this asymptotic result, we will see that the length of the confidence interval will involve the density function f which may be unknown.

In the next section we see methods of obtaining exact distribution-free confidence intervals for population quantiles, based on the probability integral transformation.

2.7   Confidence Intervals for Population Quantiles Based on Simple Random Sampling

Let ξ_p be the quantile of order p $(0 < p < 1)$ of the continuous distribution function F. Intuitively speaking it is the number such that a percentage $100p$ of the total population is numerically below or up to ξ_p. Rigorously it is defined as

$$\xi_p = \inf\{x : F(x) \geq p\}$$

leading to a unique value. $\xi_{1/2}$ is the median, the well known measure of central tendency, $\xi_{1/4}$ and $\xi_{3/4}$ are the lower and upper quartiles of the distribution.

Let $X_{(r)}$ and $X_{(s)}$ $(r < s)$ be two order statistics. Then

$$P[X_{(r)} \leq \xi_p < X_{(s)}] = P[X_{(r)} \leq \xi_p] - P[X_{(s)} \leq \xi_p]$$

$$= \sum_{i=r}^{n} \binom{n}{i} [F(\xi_p)]^i [1 - F(\xi_p)]^{n-i}$$

$$- \sum_{i=s}^{n} \binom{n}{i} [F(\xi_p)]^i [1 - F(\xi_p)]^{n-i}.$$

$$(2.7.27)$$

Assuming continuity, so that $F(\xi_p) = p$, we get

$$P[X_{(r)} \leq \xi_p < X_{(s)}] = \sum_{i=r}^{s-1} \binom{n}{i} p^i (1-p)^{n-i}.$$

The last expression is totally free of the unknown c.d.f. $F(x)$, the unknown distribution function and depends only on p, the order of the quantile for which we are seeking the confidence interval.

Hence by choosing the values of r and s appropriately, one can adjust the confidence coefficient as close as possible to $1 - \alpha$. Typically we choose r and s such that

$$\sum_{i=0}^{r-1} \binom{n}{i} p^i (1-p)^{n-i} \approx \sum_{i=s}^{n} \binom{n}{i} p^i (1-p)^{n-i} \approx \frac{\alpha}{2},$$

to obtain 'symmetric' confidence intervals. For the median, $p = 1/2$ the symmetric confidence interval will be $(X_{(r)}, X_{(n-r+1)}), 1 \leq r \leq (\frac{n+1}{2})$ and would have confidence coefficient

$$2 \sum_{i=0}^{r-1} \binom{n}{i} (\frac{1}{2})^i.$$

For example, if the sample size $n = 20$, and we take $r = 6$, then

$$2 \sum_{i=0}^{5} \binom{20}{i} (\frac{1}{2})^i \approx 0.0414.$$

This leads to $(X_{(6)}, X_{(15)})$ as the confidence interval with confidence coefficient $1 - \alpha = 0.9586$.

It is also possible to choose upper or lower confidence intervals for ξ_p by considering events of the types $\{X_{(r)} \leq \xi_p\}$ and $\{\xi_p < X_{(s)}\}$ in an analogous manner.

2.8   Confidence Intervals of the Population Quantiles Based on Ranked Set Sampling

Ranked set sampling was discussed in Chapter 1. Suppose $X_{[1]}, X_{[2]}, \ldots,$ $X_{[n]}$ is a collection of independent order statistics from n disjoint collections of n simple random samples. This constitutes a sample under ranked set sampling. The marginal distribution of $X_{[i]}$ is that of the ith order statistic from a random sample of size n. But $X_{[1]}, X_{[2]}, \ldots, X_{[n]}$ are independent. This is a collection of independent but not identically distributed random variables. Then, for $r < s$, suppose $[X_{[r]}, X_{[s]}]$ is a confidence interval for ξ_p. Since $X_{[r]}$ and $X_{[s]}$ are independent, the confidence coefficient is given by

$$
\begin{aligned}
&P[X_{[r]} \leq \xi_p \leq X_{[s]}] \\
&= P[X_{[r]} \leq \xi_p][1 - P[X_{[s]} \leq \xi_p]] \\
&= \{ \sum_{j=r}^{n} \binom{n}{j} p^j (1-p)^{n-j} \} \\
&\quad \{1 - \sum_{j=s}^{n} \binom{n}{j} p^j (1-p)^{n-j} \}.
\end{aligned} \tag{2.8.28}
$$

One notes two properties here.

(i) $E(X_{[s]} - X_{[r]}) = E(X_{(s)} - X_{(r)})$. That is the expected length of the confidence interval based on ranked set sampling is the same as the expected length of the confidence interval based on simple random sampling.

(ii) If r and s are chosen such that the confidence coefficient for the traditional confidence interval based on simple random sampling is $1 - \alpha$ with probability $\frac{\alpha}{2}$ in each tail then the confidence coefficient of the confidence interval based on ranked set sampling is

$$
(1 - \frac{\alpha}{2})^2 = 1 - \alpha + \frac{\alpha^2}{4},
$$

which increases the coverage probability by $\frac{\alpha^2}{4}$.

One should note that $X_{[r]}$ and $X_{[s]}$, being order statistics from independent simple random samples, may not satisfy the condition that $(X_{[r]} < X_{[s]})$. In some cases one may not get a proper confidence interval. In this case one could order the two statistics $X_{[r]}$ and $X_{[s]}$ as $X_{(rs1)}$ and

$X_{(rs2)}$ and use $[X_{(rs1)}, X_{(rs2)}]$ as the confidence interval. The confidence coefficient of the modified interval is

$$P[X_{(rs1)} \leq \xi_p \leq X_{(rs2)}]$$
$$= P[\{X_{[r]} \leq \xi_p \leq X_{[s]}\} \text{or} \{X_{[s]} \leq \xi_p \leq X_{[r]}\}]$$
$$= P[X_{[r]} \leq \xi_p \leq X_{[s]}] + P[X_{[s]} \leq \xi_p \leq X_{[r]}]$$
$$= P[X_{[r]} \leq \xi_p][P[\xi_p \leq X_{[s]}]]$$
$$+ P[X_{[s]} \leq \xi_p][P[\xi_p \leq X_{[r]}]]. \qquad (2.8.29)$$

Again, if r and s are chosen such that the confidence coefficient for the traditional confidence interval based on simple random sampling is $1 - \alpha$ with probability $\frac{\alpha}{2}$ in each tail, then the coverage probability of $[X_{(rs1)}, X_{(rs2)}]$ is

$$(1 - \frac{\alpha}{2})^2 + (\frac{\alpha}{2})^2 = 1 - \alpha + \frac{\alpha^2}{2}.$$

This confidence interval adds a further $\frac{\alpha^2}{4}$ to the confidence coefficient of the confidence interval based on simple random sampling. However, this comes at the cost of some increase in the expected length of the new confidence interval $[X_{(rs1)}, X_{(rs2)}]$. It can be seen that the expected length of this interval is

$$E[X_{(rs2)} - X_{(rs1)}]$$
$$= E_{F_{[s]}}[2X F_{[r]}(X) - X] + E_{F_{[r]}}[2X F_{[s]}(X) - X], \qquad (2.8.30)$$

where $F_{[s]}$ is the c.d.f. of $X_{[s]}$.

To increase the flexibility of the confidence intervals one could take m independent groups of n^2 observations for ranking purposes. From these one could obtain

$$X_{[i]j}, \quad j = 1, 2, \ldots, m, \quad i = 1, 2, \ldots, n.$$

These are m replicates of the order statistics. These $N = nm$ independent order statistics be ordered from the lowest to the highest as

$$X_{(1)} \leq X_{(2)} \leq \ldots \leq X_{(N)}.$$

Then $1 \leq r < s \leq N$ can be chosen appropriately to obtain $100(1 - \alpha)\%$ confidence interval for the pth quantile ξ_p. For details see [Öztürk and Deshpande (2004)]. Similar ranked set sampling based confidence intervals for quantiles of finite populations have been discussed in [Öztürk and Deshpande (2006)].

2.9 Tolerance Intervals

The tolerance interval is an interval (U, L) with random endpoints such that it contains at least a given proportion $1 - \gamma$ of the population with probability not less than $1 - \alpha$.

We again exploit the probability integral transformation. $(X_{(r)}, X_{(s)})$ for $r < s$ is a random interval. We can write down the following probability

$$P[F(X_{(s)}) - F(X_{(r)}) > 1 - \gamma] = P[U_{(s)} - U_{(r)} > 1 - \gamma],$$

where $U_{(s)}$ and $U_{(r)}$ are the r-th and s-th order statistics from the $U(0, 1)$ population. Their joint density function is given by

$$f_{U_{(r)}, U_{(s)}}(x, y) = \frac{n!}{(r-1)!(s-r-1)!(n-s)!} x^{r-1}(y-x)^{s-r-1}(1-y)^{n-r},$$
$$0 < x < y < 1.$$

From the above it is easy to derive the density of $U_{(s)} - U_{(r)} = V$ as

$$g(v) = n \binom{n-1}{s-r-1} v^{s-r-1}(1-v)^{n-s+r}, \quad 0 < v < 1$$

giving us

$$P[X_{(s)} - X_{(r)} > 1 - \gamma] = \int_{1-\gamma}^{1} g(v)dv,$$

which can be evaluated for given values of γ. Thus by choosing r and s appropriately the probability can be made greater than or equal to $1 - \alpha$, the required tolerance coefficient.

Thus, by choosing r and s appropriately, we obtain $(X_{(r)}, X_{(s)})$ as the interval which contains at leat $1 - \gamma$ proportion of the population with at least probability $1 - \alpha$.

Comments:

(i) The tolerance interval depends on r and s through $s - r$. Hence, the tolerance intervals are not unique.

(ii) Usually, it is possible to answer the question: What is the smallest sample size such that the sample range $(X_{(1)}, X_{(n)})$ will contain at least a proportion $(1 - \gamma)$ of the total population with probability no less than $1 - \alpha$? It would be the smallest n which satisfies the inequality

$$P[F(X_{(n)}) - F(X_{(1)}) \geq 1 - \gamma] \geq 1 - \alpha.$$

For further details on order statistics we refer to the books [David and Nagaraja (1981)] and [Arnold *et al.* (1992)].

2.10 Exercises

(1) Let $X_{(1)}, X_{(2)}, \cdots, X_{(10)}$ be the order statistics based on 10 independent identically distributed observations from a continuous distribution with $80th$ percentile 20. (i) Determine $P(X_{(8)} < 20)$. (ii) Determine $P(X_{(6)} < 20 < X_{(9)})$.

(2) Let $X_{(1)}, X_{(2)}, \cdots, X_{(7)}$ be the order statistics based on 7 independent identically distributed observations from a continuous distribution with $70th$ percentile 22. (i) Determine $P(X_{(6)} < 22)$. (ii) Determine $P(X_{(6)} < 22 < X_{(7)})$.

(3) Let $X_{(1)}, X_{(2)}, \cdots, X_{(n)}$ be the order statistics based on a random sample of size n from a continuous distribution F. Show that the distribution of $(X_{(1)}, X_{(2)}, \cdots, X_{(n)})$ is same as the distribution of $(F^{-1}(U_{(1)}), F^{-1}(U_{(2)}), \cdots, F^{-1}(U_{(n)}))$, where $U_{(1)}, \cdots, U_{(n)}$ is the order statistics based on a random sample from $U(0,1)$.

(4) Let $X_{(1)}, X_{(2)}, \cdots, X_{(n)}$ be the order statistics based on a random sample of size n from a continuous distribution F. Determine the conditional hazard rate of $X_{(j+1)}$ given $X_{(j)} = x$ in terms of the hazard rate $r_F(x) = f(x)/(1 - F(x))$ of F.

(5) Consider a k-out-of-m system, that is a system with m components which works as long as k out of the m components are working. Suppose the component lifetimes are independent and identically distributed with a common continuous distribution F. Express the distribution of the system lifetime in terms of F.

Chapter 3

EMPIRICAL DISTRIBUTION FUNCTION

3.1 Inference in the Nonparametric Setting

Let us denote by $F(x)$ the c.d.f. of the random variable X representing the random outcomes of an experiment. The approach of parametric inference is to suggest, on prior grounds, a family $\mathcal{F} = \{F_\theta, \theta \in \Theta\}$ of distribution functions. Members of this family are identified by the value of the parameter θ which is either real or vector valued (finite dimensional). Whenever θ is unknown it is estimated or hypothesis regarding it are tested on the basis of data collected by repeating the random experiment several times. For example, one may assume normal, exponential, binomial, Poisson, etc. families as the circumstances suggest. However, we are often faced with situations where such a family is hard to specify. If the experiment consists of measuring the blood pressure of a number of subjects then it may be difficult to say that random outcomes are governed by a probability distribution belonging to some such specific family. Therefore we adopt estimation and testing methods whose properties do not depend upon any particular family.

The statistical problems are now estimation of the entire distribution function F, or its values $F(x)$ at a specific argument x, or on the other hand, quantiles, i.e., the argument x where $F(x)$ takes specific values. One would like to test hypotheses regarding these entities also. We have seen how to estimate quantiles (both point and interval estimation) through the probability integral transformation in the previous chapter. In this chapter, we consider statistical inference for the entire distribution function. The empirical distribution function, nonparametric estimators arising in survival analysis and the bootstrap estimator of the distribution function are discussed in this chapter. However, the nonparametric Bayesian estimator of the distribution function is discussed in Chapter 12.

3.2   The Empirical Distribution Function

By repeating the experiment n times independently under the same conditions, one obtains the realization $x_1, x_2, \ldots, x_n$ of a random sample $X_1, X_2, \ldots, X_n$, from the distribution function under investigation, say $F(x)$. Its values at the real x may be estimated in the following way.

Define the empirical distribution function (e.d.f)

$$F_n(x) = \frac{1}{n} \sum_{i=1}^{n} I[X_i \leq x].$$

Essentially it estimates $F(x) = P[X \leq x]$ by the relative frequency of the event $[X_i \leq x, i = 1, 2, \ldots, n]$ in the random sample. If the order statistics of the random sample are $X_{(1)} \leq X_{(2)} \leq \ldots \leq X_{(n)}$ and its sample realization is $x_{(1)} \leq x_{(2)} \leq \ldots \leq x_{(n)}$, then an equivalent representation of e.d.f. is given by

$$F_n(x) = \begin{cases} 0, & \text{if } x < X_{(1)}, \\ \frac{i}{n} & \text{if } X_{(i)} \leq x < X_{(i+1)}, \quad i = 1, \ldots, n-1 \\ 1 & \text{if } X_{(n)} \leq x. \end{cases} \qquad (3.2.1)$$

A particular estimate $F_n(x)$ is illustrated in the Figure 3.1.

It is a jump function, with each jump equal to $1/n$ and located at the n order statistics $(X_{(1)}, X_{(2)}, \ldots, X_{(n)})$. Thus $F_n(x)$ will always yield a discrete (right continuous) distribution function giving probability $1/n$ to each of the order statistics. In case of ties, the appropriate adjustment to the jump size at the tied observations will be made, more specifically if k observations are tied, the jump size is taken to be k/n. The distribution function $F(x)$, which is being estimated, may or may not be discrete. However, we shall see later that in all cases $F_n(x)$ tends to be closer and closer to $F(x)$ at all x, with probability 1 as n, the sample size, becomes larger and larger. Hence it is a very attractive estimator of $F(x)$.

3.3   Properties of the Empirical Distribution Function

The empirical distribution function $F_n(x)$ is an unbiased and a weakly consistent estimator of the unknown distribution function $F(x)$.

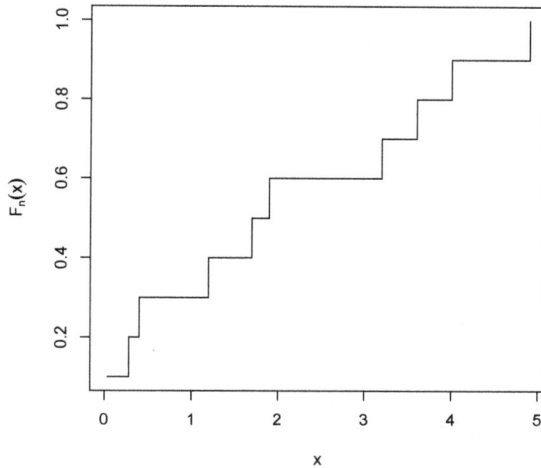

Fig. 3.1 Empirical Distribution Function

It is easy to see that for a fixed x

$$E[F_n(x)] = \frac{1}{n} \sum_{i=1}^{n} E[I[X_i \leq x]]$$
$$= P[X \leq x]$$
$$= F(x). \tag{3.3.1}$$

Hence $F_n(x)$ is unbiased for $F(x)$. It is also weakly consistent, since

$$Var(F_n(x)) = Var(\frac{1}{n} \sum_{i=1}^{n} I[X_i \leq x])$$
$$= \frac{1}{n^2} n Var I[X \leq x]$$
$$= \frac{F(x)(1 - F(x))}{n}$$
$$\to 0 \text{ as } n \to \infty. \tag{3.3.2}$$

The Borel strong law of large numbers also applies, giving $F_n(x) \to F(x)$ as $n \to \infty$ with probability 1 at fixed x.

The following theorem shows that the empirical distribution function $F_n(x)$ is a uniformly strongly consistent estimator of the unknown distribution function $F(x)$.

Theorem 3.3.1 (Glivenko-Cantelli Theorem): *Let* $X_1, X_2, \ldots, X_n$ *be i.i.d. random variables from distribution* $F(x)$. *Let* $F_n(x)$ *be the corresponding empirical distribution function. Then*

$$P[\sup_{-\infty < x < \infty} |F_n(x) - F(x)| \to 0 \text{ as } n \to \infty] = 1.$$

Proof (outline): Let $j = 1, 2, \ldots, k$ and $k = 1, 2, \ldots$ Let x_{kk} be ∞. Define x_{jk} to be the largest value of x such that

$$F(x - 0) \le \frac{j}{k} \le F(x).$$

Thus for every k, the points $x_{1k}, x_{2k}, \ldots, x_{k-1k}$ provides a partition of the real line given by $(-\infty, x_{1k}], (x_{1k}, x_{2k}], \ldots, (x_{k-1k}, \infty)$. The convergence with probability 1 of $F_n(x)$ to $F(x)$ at each end point of the above interval follows from the Borel strong law of large numbers. This is so because $F(x)$ is the probability of the event $[X \le x]$ and $F_n(x)$ is the relative frequency of this event in independent trials. Then by elementary rules of intersections and unions of events (of probability 1) we get the uniform convergence of $F_n(x)$ to $F(x)$ at all the endpoints of the above intervals with probability 1. Using the nondecreasing nature of the functions $F_n(x)$ and $F(x)$ and the definition of x_{jk} it is assured that the absolute difference $|F_n(x) - F(x)|$ for x within any of the above intervals is not more than the absolute difference at one of the endpoints of one of the interval plus $1/k$. Since k can be arbitrarily large, we can chose it so as to make $1/k$ as small as we please ensuring the uniform convergence over $(-\infty, \infty)$ of $F_n(x)$ to $F(x)$ with probability 1. See ([Loève (1963)], pp 20-21) for details.

$$\square$$

This theorem shows that $F_n(x)$, $-\infty < x < \infty$ is a good estimator of the true distribution function $F(x)$, especially if the number of observations is not too small. The functionals (parameters) of the true distribution function may then be estimated by the corresponding functionals of $F_n(x)$. For example the mean of F may be estimated by the mean of $F_n(x)$ which turns out to be $\bar{X}$, the sample mean. This is a point estimator.

We have seen that $nF_n(x) = \sum_{i=1}^{n} I[X_i \le x]$. This can be interpreted as the number of successes in n independent trials with probability of success equal to $F(x)$ at each trial. Hence, for a fixed x, $nF_n(x)$ has a *Binomial*$(n, F(x))$ distribution.

Therefore, from De-Moivre Laplace Theorem it follows that

$$\frac{nF_n(x) - nF(x)}{\sqrt{nF_n(x)(1 - F_n(x))}} \to N(0, 1) \text{ as } n \to \infty. \tag{3.3.3}$$

3.4   The M.L.E. of the Distribution Function

When we wish to carry out inference in nonparametric framework, our class of possible distributions is $\mathbb{F}$, the class of all distributions, members of which cannot be identified by a real or finite dimensional vector valued parameter. We can however, provide a justification for the empirical distribution function as a generalized maximum likelihood estimation procedure for the distribution function F itself as follows.

Among the regularity conditions for the existence of M.L.E., there is one which states that all the members of the class $\mathbb{F}$ should be absolutely continuous with respect to a common measure so that the corresponding Radon-Nikodym derivatives with respect to this measure will define the densities f. However this is impossible if $\mathbb{F}$ consists of *all* the probability distribution functions as we now want. [Kiefer and Wolfowitz (1956)] have generalized the concept in the following manner. Let $\mathbb{P} = \{P\}$ be an arbitrary class of probability measures and let P_1 and P_2 belong to it. Let

$$f(x, P_1, P_2) = \frac{dP_1(x)}{d(P_1 + P_2)(x)}$$

be the Radon-Nikodym derivative of P_1 with respect to $P_1 + P_2$. Then $\hat{P}$ is defined to be generalized maximum likelihood estimator (G.M.L.E.) of the distribution of $\underline{X}$ if

$$f(\underline{x}, \hat{P}, P) \geq f(\underline{x}, , P, \hat{P}) \ \forall P \in \mathbb{P}. \tag{3.4.1}$$

Note that if $\hat{P}$ gives positive probability to $\underline{x}$ then $f(\underline{x}, P, \hat{P}) = 0$ unless P also gives positive probability to $\underline{x}$. So we look only among those P which have $P(\underline{x}) > 0$. Then the inequality (3.4.1) reduces to $\hat{P}(\underline{x}) \geq P(\underline{x}) \ \forall P \in \mathbb{P}$.

Say, our data consists of n independent observations $x_1, x_2, \ldots, x_n$. Let an arbitrary probability distribution give positive probabilities $p_1, p_2, \ldots, p_n$ to these points. Then $P(\underline{x}) = \prod_{i=1}^{n} p_i$. It is obvious that this is maximized for $\hat{P}$ which has $p_i = 1/n$, $i = 1, 2, \ldots, n$. Hence the empirical distribution function which gives exactly probability $1/n$ to each of the observed values of the random variable is G.M.L.E. of the true distribution function in the sense of [Kiefer and Wolfowitz (1956)].

3.5   Confidence Intervals for the Distribution Function

We first look at a confidence interval for the unknown distribution function $F(x)$ for a fixed value of x. We know from (3.3.3) that for large n the standardized version,

$$\frac{nF_n(x) - nF(x)}{\sqrt{nF_n(x)(1 - F_n(x))}}$$

has $N(0,1)$ distribution. Then, for a given x, this would lead to asymptotic $(1 - \alpha)100\%$ confidence interval for $F(x)$

$$F_n(x) \pm z_{1-\frac{\alpha}{2}} \sqrt{\frac{F_n(x)(1 - F_n(x))}{n}},$$

where $z_{1-\frac{\alpha}{2}}$ is the $(1 - \frac{\alpha}{2})100\%$ upper critical point of the standard normal distribution. These confidence intervals for $F(x)$ do not need the knowledge of the family to which F belongs but are not exact. Since $F(x)(1 - F(x))$ or $F_n(x)(1 - F_n(x))$ can not exceed $1/4$

$$F_n(x) \pm z_{1-\frac{\alpha}{2}} 0.5(n)^{-1/2}$$

will provide a conservative confidence interval, meaning the confidence co-efficient will not be less than $(1 - \alpha)100\%$ whatever be the value of x. The 95% confidence interval will be

$$[F_n(x) - 0.98 \ n^{-1/2}, F_n(x) + 0.98 \ n^{-1/2}]$$

as $z_{1-\frac{\alpha}{2}} = 1.96$.

Confidence bands for the entire distribution function and tests for good-ness of fit based on empirical distribution are discussed in the next chapter.

3.6   Actuarial Estimator of the Survival Function

Here we look at the actuarial estimator of the survival function $S(x) = \bar{F}(x) = 1 - F(x)$. In demographic studies large data sets are available on the mortality experience of a population. Suppose the study begins with a cohort (sample size) of size n at time 0. The subjects are observed at fixed time points

$$0 = t_0 < t_1 < t_2 < \ldots < t_k.$$

Let, for $i = 1, 2, \ldots, k$,

$n_i =$ the number in the risk set (i.e. in the study) at time t_i,

$d_i =$ the number of deaths in the ith interval ,

$w_i =$ the number of subjects otherwise lost to the study.   (3.6.1)

Then, $n_{i-1} - n_i = d_i + w_i$ is the number of subjects which either died (failed) or were lost or withdrawn for other reasons from the study in the *ith* interval. As an approximation it is assumed that the w_i withdrawals took place uniformly over the *ith* interval or in other words only $n'_{i-1} = n_{i-1} - w_i/2$ subjects were really in the *ith* risk set.

Hence we estimate the conditional probability of a subject failing within the interval given that it had not failed at the beginning of the interval by

$$\frac{d_i}{n'_{i-1}}.$$

Using the chain rule for conditional probabilities, the probability of surviving beyond t_j is represented as

$$P[T > t_j] = \prod_{i=1}^{j} P[T > t_i | T > t_{i-1}].$$

Hence the actuarial estimator of the survival function $S(x)$ is given by

$$\hat{S}_{ACT}(t_j) = \hat{P}[T > t_j] = \prod_{i=1}^{j} (1 - \frac{d_i}{n'_{i-1}}).$$

Strictly speaking the estimator of the survival function is calculated only at the points $t_1, t_2, ..., t_k$, but we extend the domain of definition over the points in between also by attributing to them the value at the left end-point of the interval in which they are situated. However, it should be noted that the estimation is appropriate only at the end points of the intervals. The intervals are not data dependent, hence they neither become shorter nor increase in number as the data increases.

The formula for the approximate variance of $\hat{S}_{ACT}(t_j)$ called the Greenwood formula, is of interest as it would help in building asymptotic confidence intervals. Note that

$$log\hat{S}_{ACT}(t_j) = \sum_{i=1}^{j} log(1 - \frac{d_i}{n'_{i-1}}).$$

Let us assume that the failure (or survival) of subjects in the risk set is independent of each other and occurs with the same probability q_i (or p_i). Then, $n'_{i-1} - d_i$ will have $B(n'_{i-1}, p_i)$ distribution with variance $n'_{i-1}p_iq_i$.

Using the delta method we get

$$Var[log\left(\frac{n'_{i-1} - d_i}{n'_{i-1}}\right)] = \frac{q_i}{n'_{i-1}p_i}.$$

Ignoring the covariance between $d_1, d_2, ..., d_k$ we get

$$Var(log\hat{S}_{ACT}(t_j)) = \sum_{i=1}^{j} \frac{q_i}{n'_{i-1}p_i}.$$

Again by the delta method, the approximate variance of $\hat{S}_{ACT}(t_j)$ is

$$Var(\hat{S}_{ACT}(t_j)) = [\hat{S}_{ACT}(t_j)]^2 \sum_{i=1}^{j} \frac{q_i}{n'_{i-1} p_i}.$$

The expression for variance of the actuarial estimator of $\hat{S}_{ACT}(t_j)$ depends upon the unknown probabilities q_i and p_i and hence may be estimated by

$$\hat{Var}(\hat{S}_{ACT}(t_j)) = [\hat{S}_{ACT}(t_j)]^2 \sum_{i=1}^{j} \frac{d_i}{n'_{i-1}(n'_{i-1} - d_i)}.$$

By appealing to the central limit theorem, the asymptotic confidence intervals for $S(t_j)$, the value of the survival function at t_j, are given by

$$\hat{S}_{ACT}(t_j) \pm z_{1-\alpha/2} \sqrt{\hat{Var}(\hat{S}_{ACT}(t_j))}.$$

Note that t_j is an end point of the *jth* initially chosen interval.

3.7   Kaplan-Meier Estimator of the Distribution Function

If the experiment which provides observations on lifetimes can not be continued until all the lifetimes are completed, that is, until all the items put on trial have failed, we get what is called the random right censored data. With every item we associate two random variables - its lifetime X and the censoring time C at which the lifetimes got censored. We assume that the lifetime and the censoring time are independent random variables and what is actually observed for the *ith* item is

$$T_i = min(X_i, C_i)$$

and

$$\delta_i = \begin{cases} 1 & \text{if } T_i \leq C_i \\ 0 & \text{if } T_i > C_i, \end{cases}$$

which means for the *ith* individual we observe T_i, the lifetime or the censoring time, whichever is smaller and also δ_i the indicator of the observation being a completed (uncensored) lifetime. We wish to use the data (T_i, δ_i), $i = 1, 2, \ldots, n$ to consistently estimate the unknown c.d.f. F of the lifetime random variable X. Let G be the c.d.f. of C, the censoring time. Then the c.d.f. of the observed random variable T is

$$P(T \leq t) = 1 - (1 - F(t))(1 - G(t)).$$

The survival function of T is

$$P(T > t) = \bar{H}(t) = \bar{F}(t)\bar{G}(t),$$

where $\bar{F}$ and $\bar{G}$ are the survival functions of X and C, respectively.

If we obtained the empirical distribution function $H_n(t)$ formed by the data $t_1, t_2, \ldots, t_n$ it will be a consistent estimator of $H(t)$ and not of $F(t)$. So we must make use of (T_i, δ_i) to remove the effect of the c.d.f. G and obtain a consistent estimator of $F(x)$ or equivalently of the survival function $\bar{F}(x)$. We take inspiration from the actuarial estimator introduced in the previous section.

Let $T_{(1)} \leq T_{(2)} \leq \ldots \leq T_{(n)}$ be the order statistics of the data on T and define $\delta_{(i)} = \delta_j$ if $T_{(i)} = T_j$. Let $R(t)$ be defined as the set of all items still not failed at time t. Then $R(t)$ is the risk set at time t. Let n_i be the number of items in the risk set $R(T_{(i)})$ and let d_i be the number of items which fail at the time $T_{(i)}$. It may be noted that for the data with only one failure at $T_{(i)}$, the indicator function $\delta_{(i)} = 1$ or 0 according to whether $T_{(i)}$ corresponds to a failure time or a censoring time.

Let $(0, T_{(1)}], (T_{(1)}, T_{(2)}], \ldots, (T_{(n-1)}, T_{(n)}]$ be the intervals $I_1, I_2, \ldots, I_n$. Define

$$P_i = P(X > T_{(i)} | X > T_{(i-1)}).$$

That is, P_i is the conditional probability that the item survives through I_i given that it is working at time $T_{(i-1)}$. Also let

$$Q_i = 1 - P_i.$$

In the absence of tied observations the estimator of Q_i is

$$\hat{Q}_i = \frac{d_i}{n_i},$$

and hence that of P_i is

$$\hat{P}_i = 1 - \frac{d_i}{n_i} = \begin{cases} 1 - \frac{1}{n_i} & \textit{if } \delta_{(i)} = 1 \\ 1 & \textit{if } \delta_{(i)} = 0. \end{cases}$$

The survival function at time t is

$$S(t) = \bar{F}(t) = P[X > t] = \prod_{\{i : T_{(i)} \leq t\}} P_i.$$

Its product limit or the Kaplan Meier (K-M) estimator is given by

$$\bar{F}_n^{KM}(t) = \prod_{\{i : T_{(i)} \leq t\}} (1 - \hat{Q}_i).$$

Hence, when there are no ties,

$$\bar{F}_n^{KM}(t) = \prod_{\{i:T_{(i)} \leq t\}} \left(1 - \frac{1}{n_i}\right)^{\delta_{(i)}} \tag{3.7.1}$$

$$= \prod_{\{i:T_{(i)} \leq t\}} \left(1 - \frac{1}{n - i + 1}\right)^{\delta_{(i)}} \tag{3.7.2}$$

$$= \prod_{\{i:T_{(i)} \leq t\}} \left(\frac{n - i}{n - i + 1}\right)^{\delta_{(i)}}. \tag{3.7.3}$$

If there are d failures tied at any observation point, the factor contributing to the product term in (3.7.3) will be

$$(1 - \frac{d}{m})$$

where m is the size of the corresponding risk set.

If the last observation is censored then $\bar{F}_n^{KM}(t)$ never approaches 0, and it may be taken as undefined beyond $T_{(n)}$.

The difference between the [Kaplan and Meier (1958)] estimator and the actuarial estimator is that for the former the end points of the intervals are the random times of failure/censoring and for the latter the end points are fixed and chosen before the experiment.

[Kaplan and Meier (1958)] showed the consistency of this estimator.

Taking a clue from the actuarial estimator and again using the delta method one can obtain the approximate formula for the estimator of the variance of $\bar{F}_n^{KM}(t)$ as

$$\hat{Var}(\bar{F}_n^{KM}(t)) = (\bar{F}_n^{KM}(t))^2 \sum_{\{i:T_{(i)} \leq t\}} \frac{\delta_{(i)}}{(n - i)(n - i + 1)},$$

when there are no ties and then corresponding modification may be made to this 'Greenwood' formula in case of ties. Most softwares (such as R) provide commands for the calculation. In R, the package 'survival' is used for the computations (see, e.g.,[Deshpande and Purohit (2016)]).

The calculation of the Kaplan-Meier estimator is illustrated in the following example.

Example 3.1 ([Deshpande and Purohit (2016)]): Recorded failure and censored data of 12 turbine vanes are given as follows:
142, 149, 320, 345+, 560, 805, 1130+, 1720, 2480+, 4210+, 5280, 6890 (+ indicating censored observations).

TABLE 3.1

j	$t_{(j)}$	$\delta_{(j)}$	n_j	$\bar{F}_n^{KM}(t_{(j)})$
1	142	1	12	0.9167
2	149	1	11	0.8334
3	320	1	10	0.7500
4	345+	0	-	0.7500
5	560	1	8	0.6563
6	805	1	7	0.5625
7	1130+	0	-	0.5625
8	1720	1	5	0.4500
9	2480+	0	-	0.4500
10	4210+	0	-	0.4500
11	5230	1	2	0.2250
12	6890	1	1	0.0000

Figure 3.2 shows the Kaplan-Meier estimator of the survivor function for the data in Example 3.1.

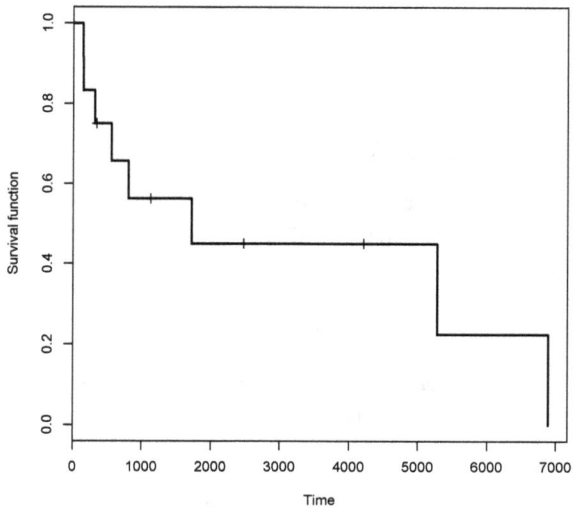

Fig. 3.2   Kaplan Meier estimate of $\bar{F}(x)$

3.8   The Nelson-Aalen Estimator of the Cumulative Hazard Function

We have discussed the Kaplan-Meier estimator of the c.d.f. in case the data is subject to random right censoring. It is a step function, with the steps located at the uncensored observations, and the height changes by a factor which is the reciprocal of the size of the risk set at that point. Then there is another estimator of the survival function called the Nelson-Aalen estimator ([Nelson (1969)], [Aalen (1975)]) which uses an indirect approach for estimating $F(x)$ through counting processes. Hence, small and large sample properties of the estimator can be derived by standard martingale theory.

Along with a continuous distribution function $F(x)$ we also have the density function $f(x)$, where

$$f(x) = \lim_{\delta \to 0} \frac{F(x+\delta) - F(x)}{\delta},$$

the probability density function (p.d.f.), which adequately describes the probability distribution of interest. In the continuous case, another function of interest is the failure (or hazard) rate function defined as

$$r(x) = \lim_{\delta \to 0} \frac{1}{\delta} P[x < X \le X + \delta | x < X]$$

$$= \lim_{\delta \to 0} \frac{1}{\delta} \frac{F(x+\delta) - F(x)}{\bar{F}(x)}$$

$$= \frac{f(x)}{\bar{F}(x)} \left(= -\frac{d}{dx} \log \bar{F}(x) \right).$$

The definition of $r(x)$ makes it clear that we seek the conditional rate of failure at time x given the survival of the item up to time x. It is also seen that

$$\bar{F}(x) = exp\{ -\int_0^x r(u) du \}.$$

Hence, there is a one to one correspondence between the distribution function and the hazard rate function. Let us define the cumulative hazard function $R(t)$ as

$$R(t) = \int_0^t r(u) du.$$

Then,

$$\bar{F}(t) = exp\{ -R(t) \}.$$

Here we seek an estimator of $R(t)$, the cumulative hazard function. Let there be n items which fail according to the c.d.f. $F(x)$, p.d.f. $f(x)$ or equivalently the failure rate $r(x)$.

Now in the framework of counting processes, define

$$Y_i(t) = \begin{cases} 1 & \text{if the } ith \text{ item is at risk just prior to time } t, \\ 0 & \text{if it has either failed or censored before time } t. \end{cases}$$

Define

$$N_i(t) = \begin{cases} 1 & \text{if the } ith \text{ item has failed or censored up to time } t, \\ 0 & \text{otherwise.} \end{cases}$$

That is to say $N_i(t)$ is the observed number of events for the ith item in the interval $(0, t]$. Define

$$N(t) = \sum_{i=1}^{n} N_i(t),$$

which counts the total number of events that have taken place in the time interval $(0, t]$ in the entire sample. Also, let

$$Y(t) = \sum_{i=1}^{n} Y_i(t),$$

that is, the total number of items still at risk just before time t. The (random) intensity process of $N_i(t)$ is assumed to be $R(t)Y_i(t)$ and that of the aggregated process $N(t)$ to be

$$\lambda(t) = \sum_{i=1}^{n} r(t)Y_i(t) = r(t)Y(t).$$

Let $T_1 < T_2 < \ldots$ be the actual failure times (not including the censoring times) of the random sample of n items. The number of failures observed is random. Without assuming any parametric form for $r(t)$ (or for $R(t)$) we can then propose

$$\hat{R}(t) = \sum_{T_i \leq t} \frac{1}{Y(T_i)},$$

as the Nelson-Aalen estimator of $R(t)$. Note that it is an increasing step function with steps at $T_1 < T_2 < \ldots$ and the addition of $1/Y(T_i)$ at the ith ordered failure time. However, this is an estimator for the cumulative hazard function $R(t)$ and not for the survival function $\bar{F}(t)$.

Alternatively, the estimator can also be written as

$$\hat{R}(t) = \int_0^t \frac{J(u)}{Y(u)} dN(u),$$

where $N(\cdot)$ is the counting process already introduced and $J(u)$ is the indicator function of the event $\{Y(u) > 0\}$. This avoids the terms with 0 in the denominator. This is a stochastic integral with respect to the counting process. It is known that the process defined by

$$M(t) = N(t) - \int_0^t r(u)Y(u)du$$

is a mean zero martingale. Let

$$R^*(t) = \int_0^t J(u)r(u)du,$$

then

$$\hat{R}(t) - R^*(t) = \int_0^t \frac{J(u)}{Y(u)}dM(u)$$

is a mean zero martingale itself. The estimated variance of $\hat{R}(t)$ is

$$\hat{\sigma}^2(t) = \sum_{T_j \leq t} \frac{1}{(Y(T_j))^2}$$

$$= \int_0^t \frac{J(u)}{(Y(u))^2}dN(u).$$

Further, as $n \to \infty$,

$$\sqrt{n}[\hat{R}(t) - R(t)]$$

converges in distribution to a mean zero Gaussian martingale with variance function

$$\sigma^2(t) = \int_0^t \frac{r(u)}{y(u)}du,$$

where $y(t)$ is a positive function such that

$$\frac{Y(t)}{n} \xrightarrow{p} y(t) \text{ for all } t \in (0, \tau] \text{ as } n \to \infty.$$

One can then obtain large sample confidence interval for $R(t)$ by using the estimated variance and the critical points from the normal distribution.

In the framework of this section, the Kaplan-Meier estimator of the survival function is

$$\bar{F}_n^{KM}(t) = \prod_{T_i \leq t} \left(1 - \frac{1}{Y(T_i)}\right),$$

and its estimated variance is

$$\hat{Var}(\bar{F}_n^{KM}(t)) = (\bar{F}_n^{KM}(t))^2 \sum_{T_i \leq t} \frac{1}{(Y(T_i))^2},$$

which is slightly different from the Greenwood expression

$$(\bar{F}_n^{KM}(t))^2 \sum_{T_i \leq t} \frac{1}{(Y(T_i))(Y(T_i) - 1)}.$$

In the above formula appropriate modifications will have to be made in case there are tied failure times.

3.9 The Nonparametric Bootstrap

Let F be an unknown c.d.f. from which the realization $x_1, \ldots, x_n$ of a random sample $X_1, \ldots, X_n$ is available. Let us consider the e.d.f. $F_n(x)$ defined in (3.2.1). The mean of the e.d.f. is

$$\hat{\mu} = \int_{-\infty}^{\infty} x\, dF_n(x) = \frac{1}{n} \sum_{i=1}^{n} x_i = \bar{x}. \tag{3.9.1}$$

It is the sample mean. Its variance is the sample variance given by

$$\hat{\sigma}^2 = \int_{-\infty}^{\infty} x^2 dF_n(x) - \{ \int_{-\infty}^{\infty} x\, dF_n(x) \}^2 = \frac{1}{n} \sum_{i=1}^{n} x_i^2 - \bar{x}^2. \tag{3.9.2}$$

The estimators $\hat{\mu}$ and $\hat{\sigma}^2$ are consistent estimators of μ and σ^2, the mean and variance corresponding to the distribution function F.

Let $W = W(X_1, \ldots, X_n)$ be a statistic based on the random sample with an unknown c.d.f. Then $G_W(w)$, the c.d.f. of W may be estimated if we have a large number (say m) of independent random samples of size n of X observations, from each of which the value of W is calculated. Let $W_1, W_2, \ldots, W_m$ denote these values. The e.d.f. obtained from $W_1, \ldots, W_m$, may be used as an estimator of the unknown c.d.f. of W.

[Efron (1979)] suggested that instead of basing the values of W on independent random samples obtained from the original distribution F, one could use random samples from the e.d.f. $F_n(x)$. Now, obtaining observations from $F_n(x)$ essentially means sampling n observations from the original sample $x_1, \ldots, x_n$, with replacement. This can be repeated a large number of times, say B. From each of these B 'new' random samples the value of the statistic W may be calculated. This is called 'Bootstrap' sampling and the collection $W_1^*, W_2^*, \ldots, W_B^*$, the bootstrap random sample of W.

The expectation and the variance of W may be estimated by

$$\hat{\mu}_W^B = \frac{1}{B} \sum_{i=1}^{B} W_i^* \quad \text{and} \quad \hat{\sigma}_W^{2\,B} = \frac{1}{B} \sum_{i=1}^{B} (W_i^* - \hat{\mu}_W^B)^2.$$

No additional experimentation is required as the 'data' consist of random samples from F_n (and not from F). The number of distinct (ordered) random samples of size n from the values $x_1, \ldots, x_n$ obtained through sampling with replacement is $\binom{2n-1}{n}$ and may be said to be equally likely.

Since the Glivenko-Cantelli theorem assures that F_n and F will be close for large n, the estimates obtained through $\hat{\mu}_W^B$ and $\hat{\sigma}_W^{2\,B}$ too are expected to be close to the expectation and variance of W.

Actually the e.d.f. $G_W^B(t)$ of the B values $W_1^*, W_2^*, \ldots, W_B^*$ may be seen as an estimator of the true c.d.f. $G_W(t)$ of W. Hence other features of $G_W(t)$, like its quantiles may also be estimated from it. One may also obtain histograms from this data to view the features of the density function of W.

If W is an unbiased estimator (maybe only asymptotically so) of a functional $\theta(F)$ of F, then the bootstrap estimator of the standard error of G_W can be used to construct approximate confidence intervals for $\theta(F)$ as long as we can obtain the values of W for a large number of random samples obtained from the e.d.f. F_n. (Thus we are able to obtain such approximate intervals without knowing either F or G_W thus qualifying as a nonparametric result.) For details we refer to [Davison and Hinkley (1997)].

3.9.1 *Bootstrap Confidence Interval Based on Normal Approximation*

Consider a general functional $\theta = \theta(F)$ of the distribution function F. It is estimated by its value $W = \theta(F_n)$ as a functional of the e.d.f. F_n. For example the mean, the variance or the third moment of $F(x)$ is estimated by the sample mean $\bar{X}$, the sample variance or the sample third moment - which are respectively the mean, the variance and the third central moment of the e.d.f. F_n.

We calculate $\hat{\sigma}_W^B$, the standard error of the bootstrap sample and suggest confidence intervals for unknown $\theta(F)$. In case the distribution of $W = \theta(F_n)$ is approximately normal with mean $\theta(F)$ then

$$(\theta(F_n) - z_{1-\alpha/2}\hat{\sigma}_W^B, \ \theta(F_n) + z_{1-\alpha/2}\hat{\sigma}_W^B)$$

may be regarded as $100(1 - \alpha)\%$ C.I. for θ, (z_β is the βth quantile of the standard normal distribution.)

3.9.2 *Bootstrap Percentile Intervals*

Let $W = \hat{\theta}_n$ be an estimator of θ based on the original data summarized by the e.d.f. F_n. If one formulates a pivotal quantity $\hat{\theta}_n - \theta$, and if its distribution were known one could, by usual inversions of inequalities, obtain a confidence interval for θ. If this c.d.f. is unknown, one can use the bootstrap method as follows.

First obtain bootstrap replications $W_1^*, W_2^*, \ldots, W_B^*$ of W. Then obtain the quantiles of order $\alpha/2$ and $1 - \alpha/2$ of the empirical distribution of W_b^*,

$b = 1, \cdots, B$, denoted by $\theta^*_{\alpha/2}$ and $\theta^*_{1-\alpha/2}$, respectively. The approximate $100(1 - \alpha)\%$ confidence interval for θ is

$$(2\hat{\theta}_n - \theta^*_{1-\alpha/2}, 2\hat{\theta}_n - \theta^*_{\alpha/2}).$$

Here the central point of the C.I is the estimator of θ obtained from the original random sample from F and the end points are based on the estimated standard errors or quantiles obtained from the bootstrap random sample.

If the original data $X_1, \ldots, X_n$ is a random sample from a c.d.f. F_θ where F is known except for the value of the parameter θ, then it is advisable to use the parametric bootstrap. It consists of first estimating θ from the data (say by the maximum likelihood method). Then obtain B bootstrap samples of size n from the distribution $F_{\hat{\theta}}$. One uses these additional bootstrap random samples to augment inference regarding θ.

Basically Bootstrap techniques work because it can be seen that in very wide setting the random variables $\sqrt{n}(F_n^B(x) - F_n(x))$ and $\sqrt{n}(F_n(x) - F(x))$ behave similarly for large n, where $F_n^B(x)$ is the e.d.f. of a bootstrap sample from F_n [Dudley (2014)] pp. 323-324.

3.10   Exercises

(1) Let F_n be the e.d.f. based on a random sample of size n from a distribution function $F(x)$. Show that

$$Cov(F_n(x), F_n(y)) = \frac{F(min(x,y)) - F(x)F(y)}{n}.$$

(2) Show that when there is no censoring the Kaplan-Meier estimator $\bar{F}_n^{KM}(t)$ reduces to the survival function corresponding to the e.d.f. (that is, the empirical survival function.)

(3) The following are failure and censoring times in years of 10 identical components, + indicates a censoring time.
    5, 7, 5+, 5.8, 11.5, 9.2, 10.2+, 11, 7+, 12+.

   (a) Using the Kaplan-Meier (KM) estimator, obtain an estimate of the probability that the component will survive beyond 8 years.
   (b) Obtain the empirical survival function by neglecting all the censored observations and compare it with the KM estimate.

(4) Is $-log(KM)$ a consistent estimator for the cumulative hazard function?

(5) Consider the data 50, 50+, 66, 70+, 85, 92, 110, 110+, 115, 118, 120+, 122

Obtain the Nelson-Aalen estimate of the cumulative hazard function $R(t)$ at $t = 100$ and at $t = 120$. Obtain large sample 95% confidence intervals for $R(100)$ and for $R(120)$.

(6) Suppose the following data are observations on a random sample of size 25 from an unknown distribution F.

$$3.7, 0.2, 1.8, 3.6, 5.4$$

$$0.5, 4.4, 0.7, 17.2, 2.5$$

$$1.8, 4.8, 2.3, 1.9, 7.5$$

$$0.6, 0.3, 5.8, 6.5, 3.2,$$

$$0.5, 0.1, 1.7, 5.0, 2.7.$$

(a) Obtain a 95% confidence interval (CI) for the mean based on the nonparametric bootstrap procedure. (Draw 1000 bootstrap samples using a computer.)

(b) Compare the above CI with the 95% large sample CI based on the normal approximation.

Chapter 4

THE GOODNESS OF FIT PROBLEM

4.1 Introduction

In Chapter 3 we discussed statistical methods for point and interval estimation of the unknown distribution function $F(x)$ for a fixed x. The estimates which we obtain are not in closed functional forms but only provide values of the function at one or all the points. Sometimes the experimenter has a suspicion or prior belief that the distribution belongs to a particular parametric family, like the Normal, exponential, Poisson etc. This could be because the experimental conditions point to a particular distribution as the appropriate one or because of past experience of similar experiments. He then wishes to either confirm or reject this prior belief through a 'test of goodness of fit'. There are three major ways of carrying out such tests:

(i) chi-squared Goodness of fit test of Karl Pearson [Pearson (1900)],

(ii) the Kolmogorov-Smirnov goodness of fit test based on the empirical distribution function [Kolmogorov (1933)] [Smirnov (1948)]), and

(iii) the Hellinger distance based methods [Beran (1977)].

We shall describe the methods (i) and (ii) in successive sections. These will be followed by methods developed for testing goodness of fit of specific popular distributions such as exponential or Normal. Method (iii) involves density estimators and will be described in the Chapter 10.

4.2 Chi-squared Goodness of Fit Test

The random sample consists of n observations $X_1, X_2, \ldots, X_n$. The idea is to see whether they occur according to a given probability distribution $F_0(x)$. The ideal situation is when we can completely specify the suspected distribution function $F_0(x)$. Often, we can only point to a particular family without being able to specify the values of its parameters. These two cases will be dealt with separately.

4.2.1   *Completely Specified Distribution Function*

We set up the following null hypothesis for testing

$$H_0 : F(x) = F_0(x) \ \forall \ x$$

against the alternative

$$H_A : F(x) \neq F_0(x)$$

over a set of nonzero probability. Choose $k > 2$ numbers

$$-\infty = a_0 < a_1 < \ldots < a_{k-1} < a_k = \infty.$$

Let $(-\infty, a_1], (a_1, a_2], \ldots, (a_{k-2}, a_{k-1}], (a_{k-1}, \infty]$ be a partition of the real line into k intervals. Since $F_0(x)$ is completely known, we can find the probabilities given by it for intervals $(-\infty, a_1], (a_1, a_2], \ldots, (a_{k-2}, a_{k-1}], (a_{k-1}, \infty]$. Let these probabilities be denoted by $p_1, p_2, \ldots, p_k$, $p_i > 0$, $i = 1, 2, \ldots, k$ and $\sum_{i=1}^{k} p_i = 1$. Let O_i be the observed number of observations in the ith interval, $\sum_{i=1}^{k} O_i = n$. The probability of the ith interval is p_i hence the expected number of observations in it is np_i. [Pearson (1900)] suggested that we should look at the discrepency between the observed and expected frequencies through the chi-squared statistics

$$\chi^2 = \sum_{i=1}^{k} \frac{(O_i - np_i)^2}{np_i}.$$

The denominator np_i is the normalizing factor to allow for unequal variances of O_i. If $F_0(x)$ is indeed the true distribution function then the difference between O_i and np_i is expected to be small, only due to random, rather than systematic variation which will arise if the probabilities p_i's are not the true probabilities. In fact, let us slightly modify the hypothesis testing problem to:

H_0' : p_i is the probability of interval $(a_{i-1}, a_i], i = 1, 2, \ldots, k$
vs
H_1' : q_i (which are not all equal to p_i) is the probability of these intervals.

Then the vector $(O_1, O_2, \ldots, O_k)$ will have a multinomial distribution under H_0' and the joint distribution is given by

$$P_{H_0}(O_1 = n_1, \ldots, O_k = n_k) = \frac{n!}{\prod_{i=1}^{k} n_i!} \prod_{i=1}^{k} p_i^{n_i},$$

and under H_1'

$$P_{H_1'}(O_1 = n_1, \ldots, O_k = n_k) = \frac{n!}{\prod_{i=1}^{n} n_i!} \prod_{i=1}^{k} q_i^{n_i}.$$

The likelihood ratio test for a simple vs. a composite hypothesis is based on the statistics

$$L = \log \frac{\sup_{H_1'} P_{H_1}}{P_{H_0'}} = \log \frac{\prod_{i=1}^{k} \left(\frac{n_i}{n}\right)^{n_i}}{\prod_{i=1}^{k} p_i^{n_i}}$$

$$= \log \prod_{i=1}^{k} \left(\frac{n_i}{np_i}\right)^{n_i}$$

$$= \sum_{i=1}^{k} n_i \left(\log \frac{n_i}{n} - \log p_i\right)$$

since $\frac{n_i}{n}$ are the maximum likelihood estimators of q_i.

By Taylor expansion, and neglecting terms of order $O(\frac{1}{n})$ we get

$$L \approx \sum_{i=1}^{k} \frac{(n_i - np_i)^2}{n_i}.$$

Replacing n_i by the quantity np_i in the denominator, which it estimates consistently, we get Pearson's chi-squared statistic. Hence, asymptotically the chi-squared statistic has the same distribution as the likelihood ratio statistic. The latter, by general principles of likelihood theory, is known to have the chi-square distribution with $k - 1$ degrees of freedom.

As large deviations between O_i and np_i, the observed and expected frequencies provide evidence against the null hypothesis, we reject it if the observed value of the chi-squared statistic is greater than the upper $\alpha\%$ value of the chi-square distribution with $k - 1$ d.f., i.e. the test is to reject H_0 if

$$\chi^2 > \chi^2_{k-1, 1-\alpha}.$$

It is clear that if there is a distribution F_1, different from F_0, but specifying the same probabilities p_i for the intervals then the test will not be effective in detecting this alternative. The construction of the intervals is rather arbitrary, it is possible that different decisions may be reached through different such constructions. The number of intervals should not be too small, but at the same time it should be kept in mind that the approximation provided by the asymptotic distribution would not be good if the probability under the null hypotheses for any of interval is too small. A rule of thumb that most statisticians recommend and follow is that n and each p_i should be large enough so that no np_i is less than 5 or so.

4.2.2   *Some Parameters of the Distribution Function are Unknown*

We have said in section (4.2.1) that the stipulated distribution function $F_0(x)$ is completely known. The experimenter sometimes may have an inkling only of the family of the distribution, but not the values of the parameters identifying the exact distribution within the family. For example, the experimenter may suspect, due to the experimental conditions, that the distribution governing the outcomes is N (μ, σ^2), but may not be able to specify, even as a hypothesis to be tested, the values of the mean μ and the variance σ^2. In such situations, it is usually suggested that the unknown scalar or vector parameter θ be estimated by its minimum chi-square estimator $\hat{\theta}$. Then the estimated value be substituted in the functional form of the distribution function $F_0(x)$ and the probabilities $\hat{p}_i, i = 1, 2, \ldots, k$ should be obtained for the k intervals. Then the statistic

$$\chi_1^2 = \sum_{i=1}^{k} \frac{(O_i - n\hat{p}_i)^2}{n\hat{p}_i}$$

can be used as before. The asymptotic distribution of the statistic χ_1^2 based on $\hat{p}_i$ is chi-square with $k - p - 1$ degrees of freedom where p is the number of parameters (dimensionality of θ) which are estimated from the data. This result again follows from the standard asymptotic theory of likelihood ratio tests. So, the critical points for the test should be chosen from the chi-square distribution with $k - p - 1$ degrees of freedom. It is thus clear that we may, at most, estimate $k - 2$ parameters from the data while testing goodness of fit.

[Pearson (1900)] developed the χ^2 test of goodness of fit of a simple (completely specified distribution) null hypotheses and found the asymptotic distribution of the statistic to be χ^2_{k-1} where k is the number of classes in which the sample space is partitioned. [Fisher (1924)] dealt with the case when the distribution is not completely specified but contains p unknown parameters. He proved that if the estimators obtained by the minimum χ^2 technique are substituted for the unknown parameters then the asymptotic distribution is χ^2_{k-p-1}. Furthermore, [Chernoff et $al.$ (1954)] showed that if estimators obtained by the more efficient maximum likelihood method of estimation then the asymptotic distribution is that of $T = \chi^2_{k-p-1} + Z^2$ where $Z^2 = \sum_{i=1}^{p} \lambda_i X_i^2$, X_i, $i = 1, 2, \ldots, p$ being independent $N(0,1)$ random variables also independent of the χ^2_{k-p-1} variable and $0 < \lambda_i < 1$. Thus, the asymptotic distribution of T is stochastically bounded between χ^2_{k-1} and χ^2_{k-p-1} random variables. In this situation using the critical points from the χ^2_{k-1} distribution will lead to a conservative test and using those from the χ^2_{k-p-1} distribution will lead to an anticonservative test, i.e., the actual level of significance will be larger than the stated one.

Example 4.1: The following are supposed to be 50 values generated from the Poisson distribution with mean 1 using a certain computer programme.

Values	0	1	2	3	4	5
frequency	11	17	10	9	2	1

Thus we wish to test the hypothesis
$H_0 : F_0$, that is, the random variable of interest is Poisson with mean 1.

i	$P_i = P[X = i]$	np_i	np_i	O_i	$(np_i - O_i)^2$	$\frac{(np_i - O_i)^2}{np_i}$
0	0.367879	18.3940	18.3940	11	54.6712	2.9722
1	0.367879	18.3940	18.3940	17	1.9432	0.1056
2	0.183940	9.1970	9.1970	10	0.6448	0.0701
3	0.061313	3.0657	3.9654	12	64.5548	16.2795
4	0.15328	0.7664				
5	0.003066	0.1533				

Entries in the third column give the expected frequencies for each value $i = 0, 1, \ldots, 5$. Since the expected frequencies corresponding to $3, 4, 5$ are too small, the entries in the fourth column represent the frequencies after the last 3 entries have been grouped. The calculated value of the χ^2 statistic is 19.427.

The upper 5% value of the chi-square distribution with $k - 1 = 3$ d.f. is $\chi^2_{3,.95} = 7.815$. Since the calculated $\chi^2 > 7.815$, we reject H_0.

The p-value in this case is less than 0.001.

Example 4.2: The data is taken from 'A Hand Book of Small Data Sets (1984), No. 181'. (Original Source: [Lieblein and Zelen (1956)]).

The number of cycles to failure of 22 ball bearings are given. The data is already in the ordered form.

17.88	28.92	33.00	41.52	52.12
45.60	48.48	51.84	51.96	54.12
55.56	67.40	68.64	68.88	84.12
93.12	98.64	105.12	105.84	127.92
128.04	173.40			

The aim is to test

$H_0 : F_0(x) = 1 - e^{-\lambda x}$, $x > 0$, $\lambda > 0$, that is, X has exponential distribution with mean $1/\lambda$. The mean is unknown.

The maximum likelihood estimator of λ is

$$\hat{\lambda} = \frac{n}{\sum_{i=1}^{n} X_i} = \frac{1}{72.3873} = 0.0138.$$

Partition	$\hat{p}_i$	$n\hat{p}_i$	O_i	$(n\hat{p}_i - O_i)^2/(n\hat{p}_i)$
(0, 40]	0.424540	9.33988	3	4.30349
(40, 80]	0.244306	5.37473	11	5.88748
(80, 120]	0.140588	3.09294	5	1.17587
(120, 160]	0.080903	1.77987	2	0.02723
(160, 200]	0.046556	1.02423	1	0.00057

In the above table $\hat{p}_i = e^{-\hat{\lambda} a_{i-1}} - e^{-\hat{\lambda} a_i}$. Thus $\chi^2 = 11.3946$.

Note that in this case χ^2 is stochastically bounded between χ^2_4 and χ^2_3. The .05 level critical points of the respective distributions are $\chi^2_{4,.95} = 9.488$ and $\chi^2_{3,.95} = 7.815$.

The null hypothesis is rejected using both the critical points.

If the observed value happens to be between the two critical points, then one should use conservative procedure, that is, use the critical value with the smaller level of significance.

4.3 The Kolmogorov-Smirnov Goodness of Fit Test

This test is based directly on the difference between the empirical distribution function $F_n(x)$ and $F_0(x)$ the distribution function specified by the null hypothesis.

Again, let the null hypothesis H_0 completely specify the distribution function:

$$H_0 : F(x) = F_0(x) \ \forall \ x$$

against the two-sided alternative H_A given in (4.2.1). $F(x)$ is assumed to be a continuous distribution function.

The random sample $X_1, X_2, \ldots, X_n$ is used to construct the empirical distribution function $F_n(x)$ defined in the Chapter 2. Calculate the Kolmogorov-Smirnov statistics

$$D_n = \sup_{-\infty < x < \infty} |F_n(x) - F_0(x)|. \tag{4.3.1}$$

The statistic D_n can be expressed as follows

$$D_n = max(D_n^+, D_n^-)$$

where

$$D_n^+ = \sup_{-\infty < x < \infty} (F_n(x) - F_0(x)), \quad D_n^- = \sup_{-\infty < x < \infty} (F_0(x) - F_n(x)).$$

For calculating the statistic D_n consider the following expressions for D_n^+ and D_n^-.

$$D_n^+ = \sup_{-\infty < x < \infty} \{ \sup_{X_{(i)} \leq x < X_{(i+1)}} (F_n(x) - F_0(x)) \}$$

$$= \max_{0 \leq i \leq n} \{ \sup_{X_{(i)} \leq x < X_{(i+1)}} (\frac{i}{n} - F_0(x)) \}$$

$$= \max_{0 \leq i \leq n} \{ \frac{i}{n} - \inf_{X_{(i)} \leq x < X_{(i+1)}} F_0(x) \}$$

$$= \max_{0 \leq i \leq n} \{ \frac{i}{n} - F_0(X_i) \}$$

$$= \max\{0, \max_{1 \leq i \leq n} \{ \frac{i}{n} - F_0(X_i) \} \}. \tag{4.3.2}$$

Similarly D_n^- can be expressed as

$$D_n^- = \sup_{-\infty < x < \infty} \{ \sup_{X_{(i)} \leq x < X_{(i+1)}} (F_0(x) - F_n(x)) \}$$

$$= \max_{0 \leq i \leq n} \{ F_0(X_{i+1}) - \frac{i}{n} \}$$

$$= \max\{0, \max_{1 \leq i \leq n} \{ F_0(X_i) - \frac{i-1}{n} \} \}. \tag{4.3.3}$$

Using (4.3.2) and (4.3.3) in (4.3.1) we get D_n is the maximum of $2n$ positive quantities.

4.3.1   *Null Distribution of the Statistic*

Under the null hypotheses $X_1, X_2, \ldots, X_n$ consist of a random sample from a continuous distribution $F_0(x)$. Because of the 1 : 1 nature of the probability integral transformation $Y = F_0(X)$, for continuous F_0, we can write

$$D_n = \sup_{0 < y < 1} |S_n(y) - y|$$

where $S_n(y)$ is the empirical distribution function of $Y_1 = F_0(X_1), \ldots,$ $Y_n = F_0(X_n)$, a random sample from the uniform $U(0, 1)$ distribution. Hence, irrespective of the functional form of the underlying distribution F, the statistic D_n will have the same distribution as the one obtained from a random sample from $U(0, 1)$ distribution. This distribution, for a given n can be found by integration but this is quite tedious. Asymptotically as $n \to \infty$, [Kolmogorov (1933)] obtained the asymptotic distribution of $\sqrt{n}D_n$ as,

$$G(z) = P[\sqrt{n}D_n \leq z]$$
$$= 1 - 2\sum_{i=1}^{\infty}(-1)^{i-1}e^{-2i^2 z^2}. \qquad (4.3.4)$$

Although it is an infinite series, it converges fast, for example $P[\sqrt{n}D_n \geq 1.36]$ is quite closely approximated by .05 when n exceeds 35.

It is easy to see that

$$E[S_n(y)] = y, \ \ 0 < y < 1,$$
$$Cov(S_n(y), S_n(z)) = \frac{y(1-z)}{n}, \ \ 0 < y < z < 1. \qquad (4.3.5)$$

Consider $0 = y_0 < y_1 < y_2 < \ldots < y_k < 1$. Then, $(nS_n(y_1), n(S_n(y_2) - nS_n(y_1)), \ldots, n(S_n(y_k) - S_n(y_{k-1})), n(1 - S_n(y_k)))$ has the multinomial distribution $MN(n, p_1, p_2, \ldots, p_k)$, where p_i is the probability that Y, a $U(0, 1)$ random variable belongs to the interval $(y_{i-1}, y_i]$. Thus by the generalized Demoivre-Laplace limit theorem, asymptotically, after normalization, this vector will have the multivariate normal distribution.

[Doob *et al.* (1949)] considered the Gaussian stochastic process $Z(t), 0 < t < 1$, with the covariance structure given by (4.3.5) and showed that the $\sup_{0<t<1} |Z(t)|$ has the Kolmogorov distribution. [Donsker (1952)] showed that the process $\sqrt{n}|S_n(y) - y|$ will asymptotically have the properties of the process $Z(t)$, hence the distribution of $\sup_{0<y<1} \sqrt{n}|S_n(y) - y|$ is the same as that of $\sup_{0<t<1} |Z(t)|$. This is an outline of the easy proof of the distribution of Kolmogorov-Smirnov statistic.

From (4.3.4) one can obtain the 100% upper critical point of the distribution. Let us call it $d_{1-\alpha}$. Then, asymptotically as $n \to \infty$,

$$P[D_n > \frac{d_{n,1-\alpha}}{\sqrt{n}}] = \alpha.$$

Under the null hypotheses $F(x) = F_0(x)$, the statistic D_n^+ has a simpler asymptotic distribution which can be derived by similar techniques. It is given by

$$P[D_n^+ > \frac{z}{\sqrt{n}}] \to 1 - e^{-2z^2} \quad \text{as } n \to \infty.$$

D_n^+ and D_n^- are identically distributed.

The exact distribution, under the null hypotheses, of D_n and D_n^+ for small sample size n is rather complicated. It has been however tabulated and exact critical points for use in testing are available.

Hence in either case (n small or large) the test is to reject H_0 in favour of the alternative H_A if

$$D_n > d_{n,1-\alpha},$$

where $d_{n,1-\alpha}$ is the upper $100\,\alpha\%$ critical point from either the exact or the asymptotic distribution of the statistic.

In case the experimenter knows that the distribution from which the data has been realized, if not F_0, falls entirely above F_0 then it is more efficient to use the statistic D_n^+.

In the opposite case, when the data, if not from F_0, is expected to be from a distribution lying entirely below F_0, one should use D_n^-.

Large values of D_n^+ and D_n^- are significant for testing the H_0 against the one-sided alternatives indicated above.

4.3.2 Confidence Bands for $F(x)$ Based on Complete Samples

Next we look at a confidence band for the entire distribution function $F(x)$. Confidence bands are random bands with the property that they envelope the entire c.d.f. from $-\infty$ to ∞ with a certain probability, whereas confidence intervals discussed earlier do this for the value of $F(x)$ at just a fixed x.

Consider the random function

$$D_n = \sup_{-\infty < x < \infty} |F_n(x) - F(x)|. \qquad (4.3.6)$$

Asymptotically as $n \to \infty$,

$$P[D_n > \frac{d_{n,1-\alpha}}{\sqrt{n}}] = \alpha.$$

That is,

$$P[\sup_{-\infty < x < \infty} |F_n(x) - F(x)| \leq \frac{d_{1-\alpha}}{\sqrt{n}}] = 1 - \alpha$$

and

$$P[F_n(x) - \frac{d_{1-\alpha}}{\sqrt{n}} \leq F(x) \leq F_n(x) + \frac{d_{1-\alpha}}{\sqrt{n}} \ \forall \ x] = 1.$$

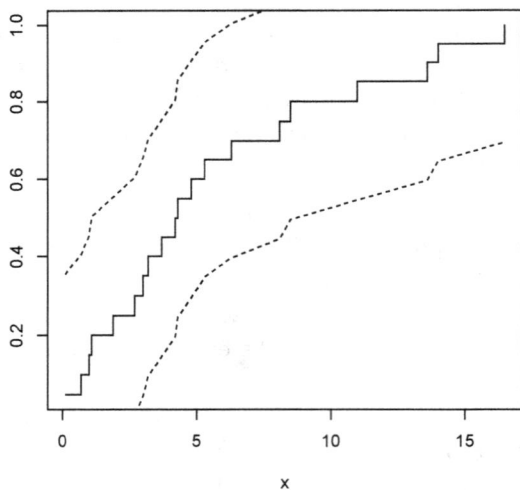

Fig. 4.1   Two-sided 95% confidence bands

The above result leads to a confidence band (given by the region enclosed by the dotted lines in Figure 4.1) for the entire unknown distribution function. The conclusion is that such bands obtained from different data sets will contain the true distribution function, entirely, in about $1 - \alpha$ proportion of times, in the long run. In a given single application, as is usual with confidence procedures, we assert that the unknown distribution is within the band, and we have a $(1 - \alpha)100\%$ 'confidence' in our assertion. If we wish to make the assertion with 95% confidence then the corresponding value of $d_{1-\alpha}$ is 1.36, provided n is at least 35. Hence the band

$$F_n(x) \pm \frac{1.36}{\sqrt{n}}, \quad -\infty < x < \infty$$

will be an appropriate one. It is interesting to compare this width with the maximum width that a 95% confidence interval for $F(x)$ at any fixed x has. We have seen in the previous section that the confidence interval for $F(x)$ is $F_n(x) \pm 0.98/\sqrt{n}$. The extra width of the band assures that the entire function $F(x)$ is contained in it with 95% confidence coefficient rather than just its value at a fixed x.

The upper 100% critical point of the distribution of D_n^+ is given by $d_{n,1-\alpha}^+$. It leads to a one sided confidence band with confidence coefficient $1 - \alpha$ as follows:

$$P[D_n^+ \geq \frac{d_{1-\alpha}^+}{\sqrt{n}}] = \alpha.$$

Therefore,

$$P[\sup_{-\infty < x < \infty} (F_n(x) - F(x)) \leq \frac{d_{1-\alpha}^+}{\sqrt{n}}] = 1 - \alpha$$

or

$$P[F(x) \geq F_n(x) - \frac{d_{1-\alpha}^+}{\sqrt{n}} \ \forall \ x] = 1 - \alpha.$$

In Figure 4.2 the broken line provides the lower bounds for the one sided confidence band for the entire unknown distribution function. In this case the 95% confidence band is obtained by using $d_{0.95}^+ = 1.22$. Note that both the two sided and one sided bands have been modified in the graphs to exclude values below 0 and above 1 which are impossible values for a distribution function in any case.

That the bands tend to be rather wide is true. But we must remember that confidence levels are truly distribution free, and the extra width is the price paid for it.

4.3.3 Confidence Bands for Survival Function Based on Censored Random Samples

In the previous chapter we have discussed the Kaplan-Meier estimator of the survival function when the data is randomly right censored. In this situation confidence bands for the survival function (or equivalently, for

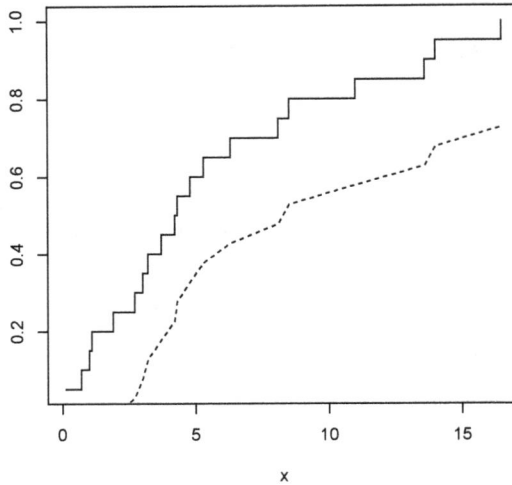

Fig. 4.2   One-sided 95% lower confidence band

the distribution function) may be formed using Hall and Wellner's results [Hall and Wellner (1980)] as follows.

Let t_{max} denote the last uncensored time. The [Hall and Wellner (1980)] asymptotic $(1-\alpha)$ level confidence bands for the survival function $\bar{F}(t)$ with $t < t_{max}$ are:

$$\left(\bar{F}_n^{KM}(t) - \frac{d_{n,(1-\alpha)}^*}{\sqrt{n}} \left[\frac{\bar{F}_n^{KM}(t)}{\bar{K}_n(t)} \right], \bar{F}_n^{KM}(t) + \frac{d_{n,(1-\alpha)}^*}{\sqrt{n}} \left[\frac{\bar{F}_n^{KM}(t)}{\bar{K}_n(t)} \right] \right),$$

where

$$\bar{K}_n(t) = \left\{ 1 + n \left(\frac{\hat{Var}(\bar{F}_n^{KM}(t))}{(\bar{F}_n^{KM}(t))^2} \right) \right\}^{-1}$$

and $d_{n,(1-\alpha)}^*$ are the critical values. For the tables with the critical values one can see [Hall and Wellner (1980)] and the book by [Klein and Moeschberger (2005)]. The package 'km.ci' in R is used for obtaining these bands.

The confidence bands in Figure 4.3 are based on the data in Table 3.1.

Confidence bands given above can be used to test for goodness of fit for a given distribution in the presence of censoring.

A general method for testing goodness of fit of a specific family of distributions $\{F_\theta, \theta \in \Theta\}$ with unknown values of parameters is to calculate

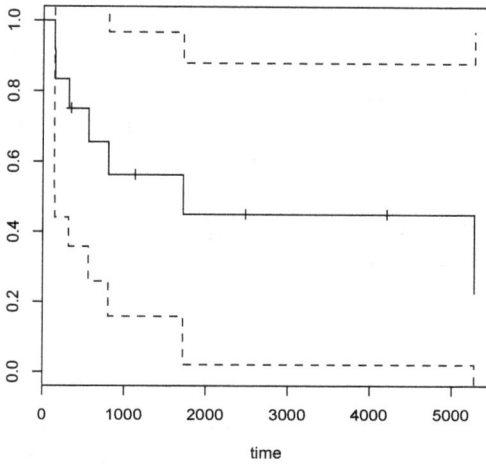

Fig. 4.3   Kaplan-Meir estimate and 90% confidence bands of $\bar{F}(x)$

$F_n(x)$ and $F_{\hat{\theta}}(x)$, where F_n is the empirical distribution function based on the data and $\hat{\theta}$ is the maximum likelihood estimator of θ in the above family. Then calculate the Kolmogorov distance

$$\hat{D}_n = \sup_x |F_n(x) - F_{\hat{\theta}}(x)|.$$

The exact (or asymptotic) null distribution of the statistic $\hat{D}_n$ will not be free of the function F or the true value θ_0 of θ. Hence critical values from the actual distribution are impractical.

If we use the critical points from the (exact or asymptotic) distribution of D_n then the test will be conservative meaning, the actual level of significance will be smaller than the stated level.

The comment made above about estimation of unknown parameters apply to modifications of one sided tests based on D_n^+ and D_n^- also.

4.3.4   *Comparison of the Chi-square and Kolmogorov Tests for the Goodness of Fit Hypotheses*

The distribution of the chi-squared statistics, under the null hypothesis, is known only asymptotically so we do not have any exact critical points for small sample sizes. Also, the test cannot distinguish the null hypothesis from another distribution which gives the same probabilities for the system of intervals. However, there is a well defined method to deal with null hypotheses which leave values of some parameters unspecified and can be applied with equal ease to continuous or discrete distributions.

In case of the Kolmogorov test, exact critical points are available for small samples also. The test is able to distinguish any distribution which is different from the distribution under the null hypothesis as long as it is continuous. However, if certain parameters are unspecified and estimated from the data then we do not know much about the error rates of the test except that it behaves in a conservative manner. Also, the distribution of the test statistic when the null hypothesis specifies a discrete distribution cannot be provided.

Hence in case of discrete distributions the chi-square test is recommended.

Tests based on Kolmogorov-Smirnov statistics can be used to test one-sided alternative hypothesis which can not be done with chi-squared statistics.

Example 4.3: Data from Example 4.2 is used to test the hypothesis
$$H_0 : F_0(x) = 1 - e^{-\lambda x}, x > 0, \lambda(> 0) \text{ unknown.}$$
The maximum likelihood estimator of λ is given by $\hat{\lambda} = 0.0138$. In the following Table $\hat{F}_0(x) = 1 - e^{-\hat{\lambda} x}$ (Mean $= 1/\hat{\lambda} = 72.3873$).

i	$x_{(i)}$	$\hat{F}_0(x_{(i)})$	i/n	$\max\{(i/n-\hat{F}_0(x_{(i)})),0\}$	$\max\{(\hat{F}_0(x_{(i)})-\frac{i-1}{n}),0\}$
1	17.88	0.218	0.045	0.000	0.218
2	28.92	0.329	0.091	0.000	0.284
3	33.00	0.366	0.136	0.000	0.275
4	41.52	0.436	0.181	0.000	0.300
5	42.12	0.441	0.227	0.000	0.259
6	45,60	0.467	0.272	0.000	0.240
7	48.48	0.488	0.318	0.000	0.215
8	51.84	0.511	0.364	0.000	0.193
9	51.96	0.512	0.409	0.000	0.148
10	54.12	0.526	0.454	0.000	0.117
11	55.56	0.535	0.500	0.000	0.081
12	67.80	0.608	0.545	0.000	0.108
13	68.64	0.612	0.591	0.000	0.067
14	68.88	0.614	0.636	0.022	0.023
15	84.12	0.687	0.682	0.000	0.051
16	93.12	0.724	0.727	0.003	0.042
17	98.64	0.744	0.772	0.029	0.017
18	105.12	0.766	0.818	0.052	0.000
19	105.84	0.768	0.864	0.095	0.000
20	127.92	0.829	0.909	0.079	0.000
21	128.04	0.829	0.954	0.125	0.000
22	173.40	0.909	1.000	0.091	0.000

Thus $D_{22} = 0.300$. Note that all entries have been approximated to third decimal place.

From the table for critical values of the Kolmogorov-Smirnov one sample test statistic we get $d_{22,0.95} = 0.281$ for the two sided test. Since $D_{22} > 0.281$ we reject H_0. The p-value in this case is 0.038. Note that we get a conservative test since a parameter has been estimated.

So far we have looked at general tests for goodness of fit. However, one could have more specific tests for goodness of fit based on statistics sensitive to departures from certain prime features of the family, such as skewness $\beta_1 = 0$ and kutosis $\beta_2 = 3$ for the Normal distribution or lack of memory property of the exponential distribution. These generally have more power for detecting departures from such features at the cost of less generality. Several such tests are studied in the following sections.

4.4 Tests of Exponentiality

The two most widely used continuous probability distributions from the modelling and applications point of view are the exponential and the Normal distributions. The exponential distribution is the single most distribution applied for modelling lifetimes. It is the only continuous distribution with the memoryless property, that is,

$$P(X > x + t | X > t) = P(X > x) \quad \forall \ x, t \geq 0. \tag{4.4.1}$$

Hence it is the proper model for the lifetimes of electronic and other non-ageing components. Also, it plays a central role in life testing as a norm, deviations from which have to be noted and studied. So it is extremely important to test goodness-of-fit of the exponential distribution to collected sets of data on lifetimes. Besides, the experimenter wishes to understand what models may be alternative to the exponential distribution. The omnibus tests like the Pearson chi-square or Kolmogorov goodness of fit tests do not provide any information on the alternatives if the null hypothesis is rejected. Hence certain tests are devised which reject the H_0 of exponentiality if certain relevant types of alternatives representing ageing hold.

As mentioned above the exponential distribution uniquely possesses the memoryless or no ageing property. But there are components which are subject to wear and tear or those which deteriorate with age. This phenomenon is known as positive ageing. One type of positive ageing is defined as follows:

$$P(X > x + t | X > t] \leq P[X > x] \quad \forall \ x, t \geq 0, \tag{4.4.2}$$

with strict inequality for some x and t.

In words we may say that a unit which has already been used for t units of time has smaller probability of surviving another x units of time compared to a new (unused) unit $\forall \ x, t \geq 0$.

A random variable X, or its c.d.f. F, which possesses the property given in (4.4.2) is said to possess New Better than Used (NBU) property.

A finer positive ageing property is the Increasing failure rate (IFR) property in which the above inequality is changed to

$$P(X > x + t_2 | X > t_2] \leq P[X > x + t_1 | X > t_1], \quad \forall \ x > 0, \ 0 < t_1 \leq t_2 < \infty. \tag{4.4.3}$$

That is, we compare two units which have already been in use for times $t_1, t_2, \ t_1 \leq t_2$, respectively. Then, the probability of both units surviving additional $x > 0$ units of times is smaller for the older unit with age t_2.

An alternative interpretation for the IFR property is that the failure rate $r(x) = f(x)/(1 - F(x))$ corresponding to random variable X is increasing in x.

There are many other classes of distributions including the Increasing failure rate average (IFRA) and the Decreasing mean residual life (DMRL) classes. A reference to any standard book of Reliability Theory, say [Barlow and Proschan (1981)], [Deshpande and Purohit (2016)] will give detailed descriptions of and interrelationships between these and such classes of distributions.

The class of Increasing Failure Rate Average (IFRA) distributions is often encountered in reliability as it is the smallest class containing the exponential distribution and closed under the formation of coherent systems. The IFRA class may be characterized by the property

$$[\overline{F}(x)]^b \leq \overline{F}(bx) \quad 0 \leq b \leq 1, \quad 0 \leq x < \infty, \tag{4.4.4}$$

with strict inequality for some b and x.

If F is IFRA, the failure rate average $\frac{1}{x} \int_0^x r(u) du$ is increasing in x.

In the following sections we discuss a few tests for exponentiality versus a few positive ageing alternatives.

4.4.1 *The Hollander-Proschan (1972) Test*

The testing problem considered in [Hollander *et al.* (1972)] is
  $H_0 : F(x) = 1 - e^{-\lambda x}, x \geq 0, \lambda > 0,$
unspecified versus the alternative
  $H_1 : \overline{F}(s + t) < \overline{F}(s)\overline{F}(t) \; \forall s, t \geq 0,$
that is, F belongs to the NBU class.

Let $X_1, X_2, \ldots, X_n$ be a random sample from the distribution F. Then the Hollander-Proschan test is based on the U-statistic estimator of the parameter

$$\gamma = \int_0^\infty \int_0^\infty \overline{F}(s + t) dF(s) dF(t)$$
$$= P[X_1 > X_2 + X_3].$$

Define a kernel function

$$h(X_1, X_2, X_3) = \begin{cases} 1 & \text{if } X_1 > X_2 + X_3, \\ 0 & \text{otherwise.} \end{cases}$$

Let $h^*(X_1, X_2, X_3)$ be its symmetrized version. Then

$$U = \frac{1}{\binom{n}{3}} \sum^* h^*(X_{i_1}, X_{i_2}, X_{i_3})$$

where $\sum^*$ is the sum over all the $\binom{n}{3}$ combinations of the indices (i_1, i_2, i_3) with $i_1 < i_2 < i_3$ from the integers $\{1, 2, \ldots, n\}$.

It is seen that $E(U) = \gamma$ which is $1/4$ under H_0 and strictly greater than $1/4$ under H_1. Also, the null asymptotic variance of $\sqrt{n}U$ is seen to be $5/432$. Hence the asymptotic distribution of

$$Z = \frac{\sqrt{n}(U - 1/4)}{\sqrt{5/432}} \;\to\; N(0, 1) \;\text{ as }\; n \to \infty.$$

The test is to reject H_0 if

$$Z > Z_{1-\alpha}$$

where $Z_{1-\alpha}$ is the $(1 - \alpha)$-th quantile of either the exact distribution of Z or its asymptotic $(N(0, 1))$ distribution.

Hollander and Proschan have shown that the test is scale invariant. It is consistent for the entire NBU class of distributions and has good efficiency for several common models belonging to this class.

4.4.2 *The Deshpande (1983) Test*

[Deshpande (1983)] studied the following testing problem
$H_0 : F(x) = 1 - e^{-\lambda x}, \quad x > 0, \quad \lambda > 0, \quad \lambda$ unknown,
versus
$H_1 : (\overline{F}(x))^b \leq [\overline{F}(bx)], \quad 0 \leq b \leq 1, \quad 0 \leq x < \infty$ and F is not exponential.

To test the null hypothesis we use the U-statistic estimator of the parameter

$$M = \int_0^\infty \overline{F}(bx) dF(x).$$

It is easily seen that $E(M) = \frac{1}{(b+1)}$ under H_0 and is strictly greater than $\frac{1}{(b+1)}$ under H_1. Hence the U-statistics

$$J_b = \frac{1}{\binom{n}{2}} \sum^* h^*(X_{i_1}, X_{i_2})$$

where $h^*(X_1, X_2)$ is the symmetric version of the kernel

$$h(X_1, X_2) = 1 \;\text{ if }\; X_1 > bX_2,$$
$$= 0 \;\text{ otherwise,}$$

and $\sum^*$ is the sum over all the $\binom{n}{2}$ combinations of (i_1, i_2) from the integers $\{1, 2, \ldots, n\}$ with $i_1 < i_2$. The asymptotic variance of $\sqrt{n}J_b$ is

$$\xi = \left\{1 + \frac{b}{2+b} + \frac{1}{2b+2} + \frac{2(b-1)}{1+b} - \frac{2b}{1+b+b^2} - \frac{4}{(b+1)^2}\right\}.$$

Then by the U-statistics theorem we know that under H_0

$$Z = \frac{\sqrt{n}(J_b - \frac{1}{b+1})}{\sqrt{\xi}} \rightarrow N(0,1) \text{ as } n \rightarrow \infty.$$

Hence the test is to reject H_0 if $Z > Z_{1-\alpha}$ where $Z_{1-\alpha}$ is again the exact $(1 - \alpha)$-th quantile of the exact null distribution or the asymptotic $N(0, 1)$ distribution of Z.

There is the question of choosing b for defining the statistic. Generally $b = 0.5$ or $b = 0.9$ is recommended. Test based on $J_{0.5}$ is consistent against the larger NBU class and $J_{0.9}$ seems to have somewhat larger power for many common IFRA distributions.

The statistics J_b is simple to compute. Multiply each observation by the chosen value of b. Arrange $X_1, X_2, \ldots, X_n$ and $bX_1, bX_2, \ldots, bX_n$ together in increasing order of magnitude. Let R_i be the rank of X_i in the combined order of these $2n$ variables. Let S be the sum of these ranks. Then it is seen that

$$J_b = \{n(n-1)\}^{-1}S.$$

It may be noted that it is essentially the Wilcoxon rank sum statistic, discussed in Chapter 7, for the data of $X_1, X_2, \ldots, X_n$ and $bX_1, bX_2, \ldots, bX_n$.

There is a hierarchy of nonparametric classes of probability distributions such as IFR, IFRA, NBU, NBUE, etc and their duals. Tests have been proposed for exponentiality against each of these classes in the literature. See [Deshpande and Purohit (2016)] for some references.

4.5 Tests for Normality

The Normal distribution is the single most commonly used model for describing the occurrence of outcomes of random experiments and phenomena. Ever since the days of Gauss and Laplace in the nineteenth century it has been recognized as a very useful model. For a considerable time it was believed that most of random phenomena actually give rise to normally distributed data, at least after appropriate transformations. Theory of errors as developed for application in Physics and Astronomy, basically makes the normality assumption. However, by and by, it came to be recognized that there are many situations where other models are much more realistically descriptive of real data. Hence there arose the need for testing whether a given set of data, i.e. realizations of independent, identically distributed random variables is described well by the Normal distribution or not. Probability plotting as explained later is a useful graphical tool in this respect. Here we describe a formal test based on the quantities involved in probability plotting.

4.5.1   *The Shapiro-Wilk-Francia-D'Agostino Tests*

Let $X_1, X_2, \ldots, X_n$ be the random sample and $X_{(1)}, X_{(2)}, \ldots, X_{(n)}$ be the corresponding order statistic. Then the test is based on the statistic

$$W = \frac{\left(\sum_{i=1}^{n} a_{i,n} X_{(i)}\right)^2}{\sum_{i=1}^{n}(X_i - \overline{X})^2}, \tag{4.5.1}$$

which is the ratio of the slope of the normal probability plot, or the square of the weighted least squares estimator of the standard deviation, to the usual estimator of the variance. The values of $a_{i,n}$ for $i = 1, 2, \ldots, n$, $n = 2, \ldots, 50$ have been tabulated. If the sample size is large say greater than 50 the following modified statistic has been proposed by [Shapiro and Francia (1972)].

$$W' = \frac{\left(\sum_{i=1}^{n} b_{i,n} X_{(i)}\right)^2}{\sum_{i=1}^{n}(X_i - \overline{X})^2 \sum_{i=1}^{n} b_{i,n}^2}$$

where $b_{i,n} = \Phi^{-1}\left(\frac{i}{n+1}\right)$ and Φ is the standard normal distribution function.

Exact critical values of W (for $n \leq 50$) and for $W'(n \leq 100)$ are available.

For even larger sample sizes [d'Agostino (1971)] proposes

$$D = \frac{\sum_{i=1}^{n}(i - \frac{1}{2}(n + 1))X_{(i)}}{n^2 s},$$

where $s = \sqrt{[\sum_{i=1}^{n}(X_i - \overline{X})^2]}$. He has provided the exact critical values for this test for small values of n.

These Shapiro-Wilk-Francia-D'Agostino tests are considered to be omnibus tests as they are able to detect departures from normality in all directions.

4.6 Diagnostic Methods for Identifying the Family of Distribution Functions

The goodness-of-fit tests described earlier in this chapter provide the means of carrying out formal statistical inference, with known probability of first type of error with respect to the distribution function governing the data.

The methods described in this section are less formal. They provide indications to the true distribution functions through graphical procedures. A suspected distribution is at the back of our mind and we compare its shape (or that of some related functions) with graphs obtained from the data.

4.6.1 *The Q-Q Plot*

The Q-Q or quantile-quantile plot compares the theoretical quantiles of a distribution with the corresponding sample quantiles represented by the order statistics. Suppose that the suspected distribution function $F(x)$ belongs to a location-scale family $F_0 \left(\frac{x-\mu}{\sigma} \right)$ where standard values of μ and σ, say 0 and 1 give a completely known standardized distribution $F_0(x)$ in this family. For example, $F(x)$ may represent the normal family with mean μ and variance σ^2, with $\mu = 0$ and $\sigma^2 = 1$ giving the standard normal distribution.

Let $F_n^*(x)$ be a slightly modified version of the empirical distribution function given by

$$F_n^*(x_{(i)}) = \frac{i - \frac{1}{2}}{n}. \tag{4.6.1}$$

Here the function takes value $(i - \frac{1}{2})/n$ rather than i/n at $x_{(i)}$. A theoretical distribution may give $-\infty$ and ∞ as the values of $F^{-1}(z)$ at $z = 0$ and 1. The modified version $F_n^*(t)$ will avoid this eventuality.

Therefore, one could compare $(F_n^*)^{-1} \left(\frac{i-1/2}{n} \right) = x_{(i)}$ and $F_0^{-1} \left(\frac{i-1/2}{n} \right)$ by plotting the points $\left(x_{(i)}, F_0^{-1} \left(\frac{i-1/2}{n} \right) \right)$ in a graph. If the true distribution $F(x)$ belongs to the location-scale family based on $F_0(x)$ then we expect that this graph called the Q-Q plot will be situated on or near a straight line. This is because

$$F_0^{-1} F(x) = F_0^{-1} F_0 \left(\frac{x - \mu}{\sigma} \right) = \frac{x - \mu}{\sigma}$$

is a straight line with slope $\frac{1}{\sigma}$ and intercept $\frac{\mu}{\sigma}$. A straight line is easy for the eye to comprehend and departures from it can be quickly recognized. While not proposing a formal test, the plot does give an indication whether the proposed location-scale family is the appropriate model or not. The slope and the intercept would provide very rough estimates of the parameters which could be useful as initial values in an iterative scheme to find, say, the maximum likelihood estimators or other more formal estimators. The values of the inverse function F_0^{-1} at the points $(i - \frac{1}{2})/n$ for $i = 1, 2, \ldots, n$, are sometimes easy to obtain by direct calculations, sometimes they are available in well known tables (e.g. Φ^{-1}, the inverse of the standard normal distribution). For many standard distributions they can be obtained by using the command 'q{distribution name}' in 'R'; for example for the normal quantiles use 'qnorm()' for the Weibull distribution use 'qweibull' etc., or can be obtained by numerical integration or other computer based calculations. Figure 4.4 shows a Q-Q plot when the data were generated from a Cauchy distribution with both location and scale parameters equal to 2 and $F_0(x)$ was taken to be the standard Cauchy distribution. Whereas Figure 4.5 shows a Q-Q plot when the data are from a Normal distribution with the mean and variance equal to 2 but F_0 was taken as the standard Cauchy distribution. In Figure 4.4, we can see that the points are situated on a straight line whereas in Figure 4.5 they are not on a straight line.

4.6.2   *The log Q-Q Plot*

This is a modification of the Q-Q plot. For some positive valued random variables the distributions of its logarithm belong to a location-scale family. For example the lognormal or the Weibull distributions have this property. Therefore, arguing as before, we plot the points $\left\{ \log x_{(i)}, F_0^{-1} \left(\frac{i - \frac{1}{2}}{n} \right) \right\}$. For example, in the Weibull case

$$F(x) = 1 - e^{-\lambda x^\nu}, \quad x > 0.$$

Hence

$$F^{-1}(y) = \frac{\log[-\log(1 - y)] - \log \lambda}{\nu}, \quad 0 < y < 1,$$

and $\lambda = \nu = 1$ leads to the standard exponential distributions with distribution function $F_0(x) = 1 - e^{-x}, x > 0$ in this family. Hence if we plot the points $(\log x_{(i)}, \log(- \log \left(1 - \frac{i - \frac{1}{2}}{n} \right)))$ they are expected to lie on a straight line with slope ν and intercept $\log \lambda$. Thus the fact that the points look like being on a straight line will indicate that the distribution is Weibull, and the slope and intercept leading to preliminary estimation of the parameters.

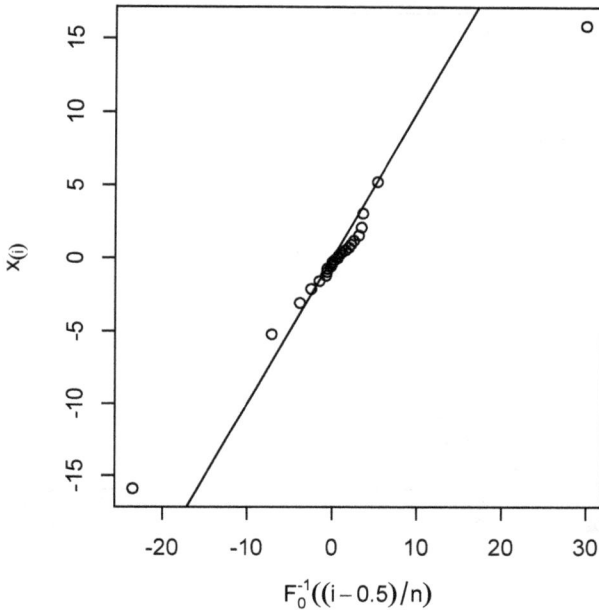

Fig. 4.4   Q-Q plot

Figure 4.6 indicates that the data are from a Weibull distribution with 2.4 and -10 as the preliminary estimates of the parameters ν and $\log \lambda$.

4.6.3   The P-P Plot

The P-P (Probability-probability) plot charts the points $(F_n^*(x_{(j)}),$ $F(x_{(j)}, \hat{\theta}))$ where F is the proposed family of distribution, possibly dependent upon parameter θ. The parameter θ may be estimated by some method suitable for this family, like the method of maximum likelihood and the estimate substituted for the true value. As before $F_n^*(x_{(j)}) = \frac{j-1/2}{n}$. This plot is restricted to the square $(0, 1) \times (0, 1)$ and the points are expected to lie on the diagonal from $(0, 0)$ to $(1, 1)$ if the model holds.

In Figure 4.7 it will be hard to say that the points do not lie on or near the diagonal, whereas in Figure 4.8 the plot seems to be concave in nature rather than the straight line of the diagonal. The shape of the graph of these points when it is not a straight line also gives some indications regarding

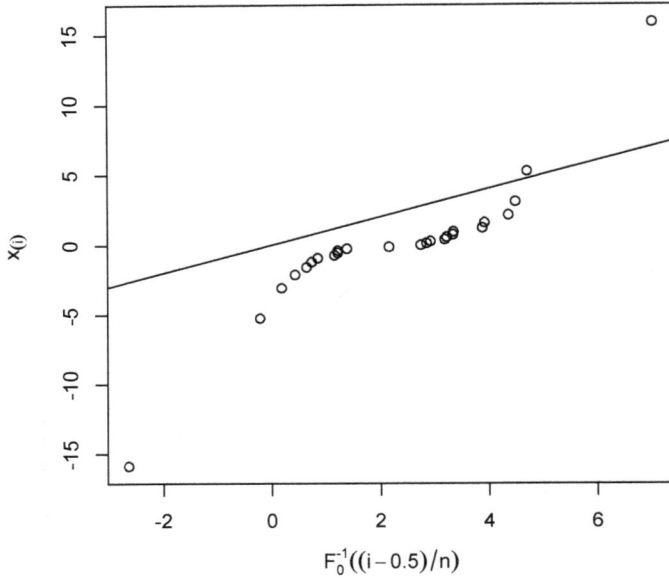

Fig. 4.5   Another Q-Q plot

the true distribution vis-a-vis the suspected distribution. In particular, if the graph is concave as in Figure 4.8, it is indicated that $r_{F_{true}}(x)/r_{F_0}(x)$ the ratio of the failure rate of the true distribution with that of the suspected distribution is increasing. This, in turn, can be interpreted to mean the data comes from a distribution which is aging faster than the suspected distribution. These considerations helps us in selecting appropriate models from the point of view of survival theory.

4.6.4   The T-T-T Plot

The total time on test (T-T-T) plot is very useful for checking adherence to the exponential model and also departures from it in specific directions which are of interest in lifetime studies. The basis is the scaled T-T-T

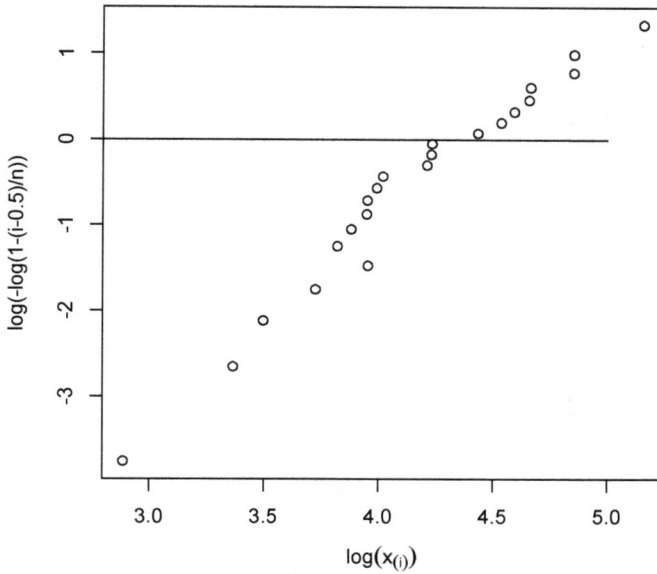

Fig. 4.6   log Q-Q plot

transform of distribution function defined by

$$T_F(u) = \frac{\int_0^{F^{-1}(u)} \overline{F}(t)dt}{\int_0^{\infty} \overline{F}(t)dt}, \quad 0 < u < 1. \tag{4.6.2}$$

Like other transforms, this is also in 1:1 correspondence with probability distributions. It is easy to see that for the exponential distribution ($\overline{F}(x) = e^{-\lambda x}, x > 0, \lambda > 0$) it is the straight line segment (diagonal) joining $(0,0)$ with $(1,1)$. Hence the technique is to define the sample version of the scaled T-T-T transform as the T-T-T statistic given by

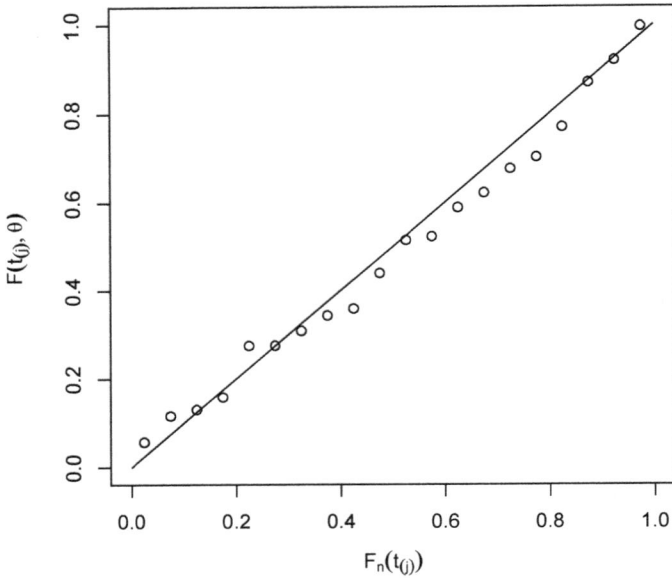

Fig. 4.7   P-P plot

$$T_{F_n}(i/n) = \frac{\int^{F_n^{-1}(i/n)} \overline{F}_n(t)dt}{\int_0^\infty \overline{F}_n(t)dt}$$

$$= \frac{\sum_{j=1}^i (n-j+1)(X_{(j)} - X_{(j-1)})}{n\overline{X}},$$

where $\overline{X}$ is the sample mean and $0 = X_{(0)} \leq X_{(1)} \leq \ldots \leq X_{(n)}$ are the order statistics of the random sample. The numerator of $T_{F_n}(i/n)$ is the total time on test (or under operation) of all the n items put on test, simultaneously, up to the i-th failure. Hence the name of the statistic and the transform. The points $\left\{\frac{i}{n}, T_{F_n}(\frac{i}{n})\right\}, i = 1, 2, \ldots, n$ are plotted in the square $(0,1) \times (0,1)$. If they lie on the diagonal or near it and not systematically on one side, then the exponential distribution is indicated.

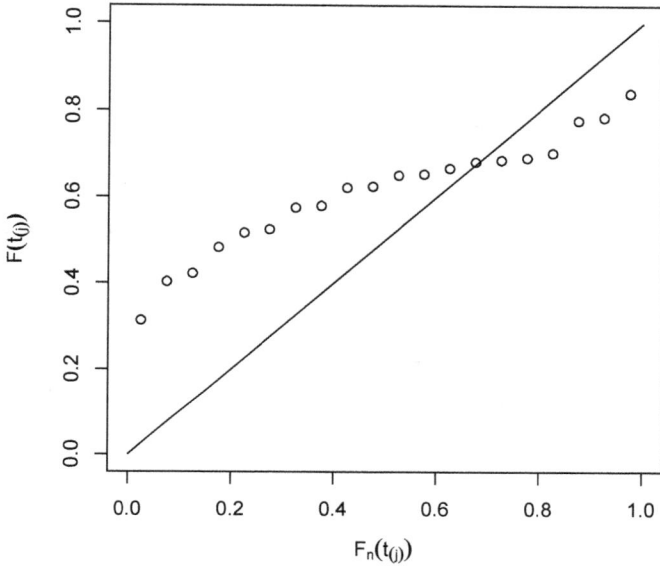

Fig. 4.8   Another P-P plot

If a systematic pattern, apart from the diagonal, is discernible then certain alternative models may be more appropriate.

In Figure 4.9 the exponential distribution is indicated, whereas in Figure 4.10 the jumps in the values of the sample scaled T-T-T transform seem to become larger and larger indicating a distribution in which failures occur progressively less frequently in time compared to the exponential distribution. The dual of the IFR class is the Decreasing Failure Rate (DFR) class and that of the NBU class is the New Worse than Used (NWU) class. If the graph appears to be convex then a DFR distribution and if it is only below the diagonal without being convex then some other NWU distribution is expected to fit better to the data.

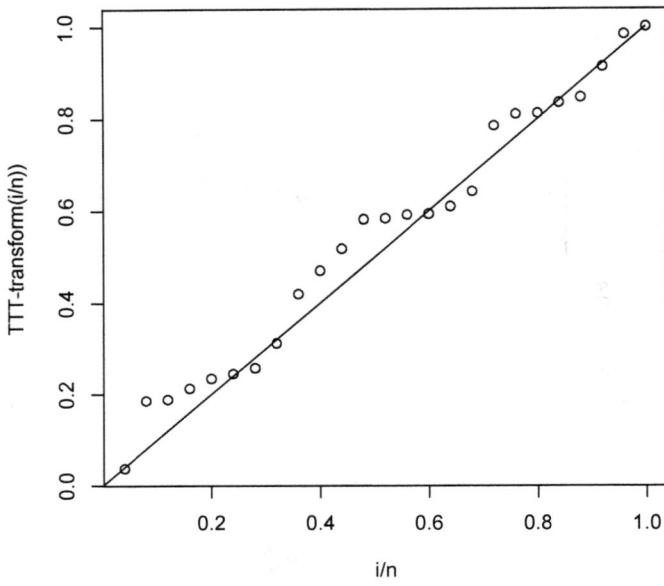

Fig. 4.9   T-T-T plot

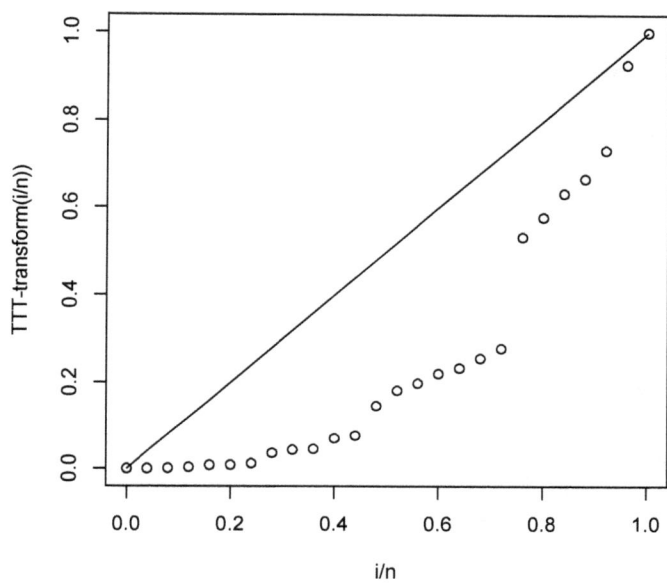

Fig. 4.10   Another T-T-T plot

Chapter 5

THE ONE SAMPLE PROBLEM

5.1 Introduction

Let us consider the situation where the same experiment is conducted n times independently. Thus the random outcomes may be represented by $Y_1, Y_2, \ldots, Y_n$, a random sample of size n. That is, there is a collection of n independent and identically distributed random variables with a common c.d.f. F. We represent the realizations of these random variables by $y_1, y_2, \ldots, y_n$, which are real numbers. One can answer many questions regarding F without knowing the parametric family to which it belongs. For example, one may want to know whether a given number θ_0 is the median of the distribution, or whether F is a symmetric probability distribution. One could also test whether a given set of observations constitute a random sample. We will take up such one sample problems one by one in this chapter. In the previous chapter we have already considered the "goodness of fit problem", which is also a one sample problem.

5.2 The Sign Test for a Specified Value of the Median

The question of interest here is whether θ_0 is the median of a continuous (or a discrete) distribution $F(x)$, to be answered on the basis of the random sample $Y_1, Y_2, \ldots, Y_n$ obtained from it. The median is defined as the point θ such that

$$\theta = \inf\{z : F(z) \geq \frac{1}{2}\}.$$

This definition uniquely defines the median θ and in case $F(z)$ is strictly monotone in an interval $[z_1, z_2]$ where $F(z_1) < \frac{1}{2}$ and $F(z_2) > \frac{1}{2}$, then the solution to $F(z) = \frac{1}{2}$ is the median. In such situations we have

$$P[Y < \theta] = P[Y > \theta] = \frac{1}{2}.$$

The median is a common and easily understood measure of the location, or the central tendency of a probability distribution and it always exists. Let us consider the null hypothesis

$$H_0 : \theta = \theta_0$$

against the alternative

$$H_1 : \theta < \theta_0, \qquad (5.2.1)$$

where θ_0 is a known number. The shift is indicated in Figure 5.1.

Equivalently, by subtracting θ_0 from each observation, one could test

$$H_0 : \theta = 0 \quad \text{vs} \quad H_1 : \theta < 0.$$

The Sign statistic is defined as

$$S^- = \sum_{i=1}^{n} I[Y_i \leq 0], \qquad (5.2.2)$$

where

$$I[Y_i \leq 0] = \begin{cases} 1 & \text{if} \quad Y_i \leq 0, \\ 0 & \text{if} \quad Y_i > 0. \end{cases}$$

Thus S^- is the number of non positive observations (or negative signs) in the random sample. Then it is seen that S^- has the Binomial distribution $B(n, F(0))$ and under H_0, $F(0) = \frac{1}{2}$, and under H_1, $F(0) > \frac{1}{2}$.

Hence an appropriate test for H_0 vs H_1 is : Reject H_0 if

$$S^- > S_{1-\alpha},$$

and the critical point $S_{1-\alpha}$ is the upper α quantile of the $B(n, \frac{1}{2})$ distribution. It is understood that the Binomial distribution is discrete, hence it may not be possible to obtain an exact $100\alpha\%$ critical point. But the distribution quickly tends to the Normal distribution. Under H_0,

$$\frac{2}{\sqrt{n}} (S^- - \frac{n}{2}) \xrightarrow{d} N(0,1) \quad \text{as} \quad n \to \infty.$$

Hence, critical points from the Normal distribution are quite accurate even for n as small as 20. The power function of the test is given by $P_{H_1}(S^- > S^-_{1-\alpha})$, which too can be obtained easily since S^-, under H_1 will have $B(n, F(0))$ distribution and $F(0) > \frac{1}{2}$ if $\theta \ (< 0)$ is the true median. The asymptotic properties too follow similarly. The standardised version

$$\frac{S^- - nF(\theta)}{[nF(\theta)(1 - F(\theta))]^{\frac{1}{2}}} \xrightarrow{d} N(0,1) \quad \text{as} \quad n \to \infty.$$

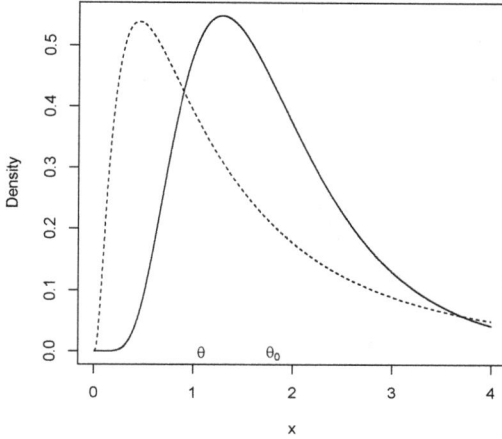

Fig. 5.1 Densities showing Shift in Median

The standardised version can be evaluated in terms of the values of $F(\theta)$ ranging from $\frac{1}{2}$ to 1 under H_1.

Example 5.1: A microwave oven is designed to emit a radiation level of 0.15 through the closed door. The following are the radiation levels of 20 ovens selected at random.

$$.09, .18, .10, .05, .12, .40, .10, .05, .03, .20,$$
$$.08, .10, .30, .20, .02, .01, .10, .08, .16, .11$$

Does the data support the hypothesis

$$H_0 : F(.15) = .5$$

against the alternative

$$H_1 : F(.15) < .5?$$

Since the alternative hypothesis is that the probability of radiation falling below .15 is less than .5, we should reject H_0 when the value of S^- is too small. Here

$$S^- = 14, \quad \text{and} \quad \frac{2}{\sqrt{n}}(S^- - \frac{n}{2}) = 1.788.$$

The critical point corresponding to the 5% or any other reasonable level α value would be negative. The observed value in this case is positive. Hence H_0 is not rejected. The p-value is .96.

The exact and asymptotic sign test can be implemented in R using the signmedian.test function in the signmedian.test package.

Comments:

(i) As the example shows the critical region is changed according to the alternative hypothesis in play.

(ii) The slight modification of the test statistic which counts the number of observations smaller (or larger) than a value θ_p, which is the pth quantile of the distribution, will be appropriate to test $H_0 : \theta_p = \theta_p^0$. If S^- is the statistic then S^- has $B(n,p)$ distribution and asymptotically $\frac{S^- - np}{\sqrt{np(1-p)}}$ will have standard Normal distribution. The critical region for a one-sided or a two-sided alternative can be chosen on considerations similar to those discussed above.

5.3   Wilcoxon Signed Rank Test

Again we present a procedure for testing the null hypothesis that the median of a distribution has the specified value θ_0. This time we make a further assumption that the probability distribution is symmetric around its median, that is, $f(\theta - x) = f(\theta + x)$, $\forall$ x. We obtain n independent observations $Y_1, Y_2, \ldots, Y_n$ from this distribution and wish to test $H_0 : \theta = \theta_0$, or equivalently, $X_i = Y_i - \theta_0$ is a random sample from a symmetric distribution with 0 as the median (or the centre of symmetry). Here we suggest the Wilcoxon signed rank test. Historically, this one sample procedure came after the corresponding two sample Wilcoxon (Mann-Whitney) test discussed in a later chapter. To compute this test statistic we require two sets of statistics from the data. First of all

$$\Psi_i = I(X_i > 0),$$

the indicator function of positive (modified) observations and

$$R_i = \text{Rank of } |X_i| \text{ among } |X_1|, |X_2|, \ldots, |X_n|, \quad i = 1, 2, \ldots, n.$$

Then we calculate

$$W^+ = \sum_{i=1}^{n} \Psi_i R_i, \qquad (5.3.1)$$

which gives the sum of the ranks of those observations which were originally greater than θ_0, in the ranking of the absolute value of all the observations, after subtracting θ_0 from each one of them.

If $\theta > 0$, then one will expect a larger proportion of the observations to be positive and also these would lie further away from 0 than those with originally negative signs. Hence for testing $H_0 : \theta = \theta_0$ vs $H_1 : \theta > \theta_0$ one would use the large values of W^+ as the critical region. On the other hand if the alternative is $H_1 : \theta < \theta_0$, then the natural rejection region will be small values of W^+.

The values of the statistic W^+ range from 0 when all Ψ_i are 0, that is, all observations are negative to $\frac{n(n+1)}{2}$ when all Ψ_i are 1, that is, all observations are positive. In fact it can be seen that under H_0 the vectors $\underline{\Psi}$ and $\underline{R}$ are independent leading to the moment generating function

$$M_{W^+}(t) = \frac{1}{2^n} \prod_{j=1}^{n} (1 + e^{tj}),$$
(5.3.2)

which by converting it into a probability generating function, can give the exact null distribution of the statistic W^+. It also gives

$$E_{H_0}(W^+) = \frac{n(n+1)}{4},$$

$$Var_{H_0}(W^+) = \frac{n(n+1)(2n+1)}{24}.$$
(5.3.3)

The asymptotic distribution of W^+, under H_0, is Normal, that is,

$$W^{+*} = \frac{W^+ - \frac{n(n+1)}{4}}{\sqrt{\frac{n(n+1)(2n+1)}{24}}} \xrightarrow{d} N(0,1), \quad \text{as } n \to \infty.$$
(5.3.4)

Thus, for large, or even moderately large sample sizes it is possible to use the critical points from the standard Normal distribution.

It is possible that the real data exhibit tied observations. Then, we assign to all the observations in the group the average of the ranks of the observations in the group. If there are s groups of tied observations with $t_1, t_2, \ldots, t_s$ observation in these groups, then the conditional variance of the statistic is reduced to

$$\frac{1}{24}[n(n+1)(2n+1) - \sum_{i=1}^{s} t_i(t_i - 1)(t_i + 1)].$$

Replacing the denominator of W^{+*} by the square root of this reduced variance will again lead to a statistic with asymptotically standard Normal distribution.

Another useful representation of the Wilcoxon signed rank statistic is

$$W^+ = \sum_{1 \le i \le j \le n} I(X_i + X_j > 0) = \sum_{1 \le i \le j \le n} T_{ij} \quad \text{(say)},$$
(5.3.5)

where I is the indicator function of the set in the paranthesis. Under the alternative hypothesis H_1 : the distribution of X is not symmetric about 0. This expression enables easy calculation of the mean and the variance of the statistic W^+ under the alternative hypothesis. Let

$$p_1 = P(X_i > 0),$$
$$p_2 = P(X_i + X_j > 0),$$
$$p_3 = P(X_i > 0, X_i + X_j > 0),$$
$$p_4 = P(X_i + X_k > 0, X_i + X_j > 0).$$

Then, some calculations lead to

$$E(T_{ii}) = p_1, \quad E(T_{ij}) = p_2.$$

Hence

$$E(W^+) = np_1 + \frac{n(n-1)}{2}p_2.$$

Besides

$$Var(W^+) = np_1(1 - p_1) + n(n-1)(n-2)(p_4 - p_2^2)$$
$$+ \frac{n(n-1)}{2}[p_2(1 - p_1) + 4(p_3 - p_1 p_2)]. \tag{5.3.6}$$

It is seen that under H_0,

$$p_1 = p_2 = \frac{1}{2}, \quad p_3 = \frac{3}{8} \text{ and } p_4 = \frac{1}{3}.$$

All this can be put together to verify that under H_0 the values obtained earlier in (5.3.3) are also obtained through this general formula also.

Example 5.2: Below are given measurements of paper density (gms/cubic cm) on 15 samples.

.816 .836 .815 .822 .822 .843 .824 .788

.782 .795 .805 .836 .738 .772 .776

It is expected that above data is a random sample from a symmetric distribution. The null hypothesis to be tested is

$$H_0 : \text{median} = .800.$$

So we subtract .800 from each observation and rank the absolute values. We note that several observations are tied together. So we assign the average rank to all observations in each tied group. The sum of ranks of the observations which were positive is

$$1.5 + 4 + 5 + 7.5 + 7.5 + 9.5 + 12.5 + 12.5 + 14 = 74.0.$$

Due to the ties the variance is reduced to $\frac{1}{24}[(15 \times 16 \times 31) - (4 \times 2 \times 1 \times 3)] = 309$. The value of the statistic is then

$$W^{+*} = \frac{74.0 - 60.0}{\sqrt{309}} = \frac{14}{17.6} \simeq .795.$$

Adopting the asymptotic standard Normal distribution, with the alternative $H_1 : \theta > 0.8$, the p-value of the observed statistic is 0.213, which is too large for rejecting H_0. So we do not do so.

In R, the function wilcox.test from the stats package implements the Wilcoxon signed rank test.

Comments:

(i) A common situation where the testing of θ as the value of the median of a symmetric distribution arises is the paired data case. Paired data arises when, among other situations, two measurements are taken on the same subject, say before and after a treatment is applied. The subjects could be patients with high fever and the measurements are made on temperature before and after the administration of a suggested drug. We consider the difference: measurement before the treatment − measurement after the treatment. The realizations of the experiment could be observations on a random variable which is symmetric with median 0 under the null hypothesis of 'no effect of drug'. Then one can apply the Wilcoxon signed rank test for testing this null hypothesis.

(ii) Even if the null hypothesis states that the distribution of the difference in the paired observations is symmetric about some nonzero value θ_0, the same procedure can be followed by subtracting θ_0 from the difference of observations.

(iii) It should be kept in mind that when rejection occurs, it may be because

(a) the symmetric distribution does not have 0 or θ_0 as the centre of symmetry, or also because

(b) the distribution is not symmetric at all, one tail of the distribution being heavier than the other.

5.4 Ranked Set Sampling Version of the Sign Test

In section 1.8 we have explained the concept of Ranked Set Sampling (RSS) methodology and its importance. The importance is mainly due to availability of auxillary information on experimental units about the relative ranks of the character in addition to the few order statistics chosen for actual measurement and further use.

Thus, k groups of k units are ranked separately and the smallest $X_{(1)}$, the second smallest $X_{(2)}, \ldots$, and the largest $X_{(k)}$ are chosen from the k groups separately. These are order statistics, but independent since they are from different groups. This process is repeated m times to yield

$$X_{(j)1}, X_{(j)2}, \ldots, X_{(j)m}, \quad j = 1, 2, \ldots, k$$

which are the independent realizations of all the k independent order statistics. Thus we have finally $n = mk$ observations. This process is adopted because it is deemed to be difficult to rank k objects (without measuring the character of interest) for a large or even moderate values of k. Typically the values of k used are $3, 4, 5$. So as not to limit the number of observations for analysis in this manner, the device of repeating the procedure m times to make $n = mk$ a reasonably large number of observations.

The observations $X_{(j)i}$, $i = 1, 2, \ldots, m$ are m independent realizations of the jth order statistic with c.d.f. $F(x - \theta)$. It would have median θ if F has the median 0.

The sign statistic based on the RSS methodology, as suggested by [Hettmansperger (1995)] for testing $H_0 : \theta = \theta_0$ is then

$$S_{RSS}^+ = \sum_{j=1}^{k} \sum_{i=1}^{m} I(X_{(j)i} > \theta_0)$$

$$= \sum_{j=1}^{k} S_j^+,$$

where $S_j^+ = \sum_{i=1}^{m} I(X_{(j)i} > \theta_0)$. It is then noted by [Hettmansperger (1995)] and by [Koti and Jogesh Babu (1996)] that S_j^+ are independently distributed as Binomial random variables with parameters $(m, 1 - I_{\frac{1}{2}}(j, k - j + 1))$. Here,

$$I_x(j, k - j + 1) = \frac{1}{\beta(j, k - j + 1)} \int_0^x u^{j-1}(1 - u)^{k-j} du,$$

and $\beta(j, k - j + 1)$ is the above integral from 0 to 1. Hence, under H_0,

$$E(S_j^+) = m(1 - I_{\frac{1}{2}}(j, k - j + 1)),$$

$$E(S_{RSS}^+) = \sum_{j=1}^{k} E(S_j^+) = \frac{mk}{2},$$

$$Var(S_{RSS}^+) = \frac{mk\lambda_0^2}{4},$$

where

$$\lambda_0^2 = 1 - \frac{4}{k} \sum_{j=1}^{k} (H_j(\theta_0) - \frac{1}{2})^2,$$

and

$$H_j(x) = \frac{k!}{(j-1)!(k-j)!} \int_{-\infty}^{x} H^{j-1}(u)(1 - H(u))^{k-j} h(u) du.$$

The c.d.f. $H(x) = F(x-\theta)$ and h(x) is its p.d.f under H_0. [Hettmansperger (1995)] has proved that the variance of S_{RSS}^+ is smaller than the variance of S^- (or S^+), the sign statistic based on simple random sampling introduced in Section 2 above. Using the basic non identical but independent Binomial random variables $S_{(i)}^+$, [Koti and Jogesh Babu (1996)] have obtained the exact, as well as the asymptotic distributions of S_{RSS}^+ under H_0. This is possible since $H(\theta_0) = \frac{1}{2}$, under H_0 and the Binomial distribution has a known probability of success. The test is to reject $H_0 : \theta = \theta_0$ against $H_1 : \theta \neq \theta_0$ if

$$S_{RSS}^+ \geq n - u \quad \text{or} \quad S_{RSS}^+ \leq u,$$

where u is an integer to be read from the tables provided by [Koti and Jogesh Babu (1996)] such that the probabilities in the tails are as close to $\frac{\alpha}{2}$ each for the required value of α.

For example, for $k = 3, m = 10$ (so that $n = 30$), the cut off points for $\alpha = .01$ (approximately) would be 9 and 21. That is,

$$P_{H_0}[S_{RSS}^+ \leq 9] = P_{H_0}[S_{RSS}^+ \geq 21] = .005.$$

Then the standardized version

$$\frac{S_{RSS}^+ - E_{H_0}(S_{RSS}^+)}{\sqrt{Var_{H_0}(S_{RSS}^+)}} \xrightarrow{d} N(0,1), \quad \text{as } n \to \infty.$$

The values of λ_0^2 are available in Table 5.1 of [Chen *et al.* (2003)]. For example, for $k = 4$ the value of λ_0^2 is 0.547. This gives an idea of the reduction of variance of S_{RSS}^+ and the increase in efficiency of the RSS procedures over the SRS procedures.

5.5 Ranked Set Sampling Version of the Wilcoxon Signed Rank Test

Next we define the Wilcoxon signed rank statistic based on ranked set sampling (RSS) data. As already specified $X_{(j)i}$, $i = 1, 2, \ldots, m$ are m independent realizations of the *jth* order statistic distribution $j = 1, 2, \ldots, k$,

$km = n$. Define

$$W_{RSS}^+ = \sum_{(r_1, i_1)} \sum_{(r_1, i_1)} I(\frac{X_{(r_1)i_1} + X_{(r_2)i_2}}{2} > 0), \qquad (5.5.1)$$

where the summation is over 1 to k for r_1 and r_2 and over 1 to m for i_1 and i_2, subject to either $i_1 < i_2$ or $r_1 \le r_2$ if $i_1 = i_2$. This statistic is also suggested for testing the $H_0 : \theta = 0$, where θ is the centre of symmetry of a continuous symmetric distribution $F(x, \theta)$. The expectation and variance of W_{RSS}^+ under H_0 can be obtained through the distribution and probability density functions of uniform order statistics. It can be verified that

$$E_{H_0}(W_{RSS}^+) = \frac{n(n+1)}{4},$$

where $n = mk$, which is the same as the expression for $E_{H_0}(W_{SRS}^+)$. The variance calculations, even under the null hypothesis are more involved. The details are given in [Chen *et al.* (2003)]. It is shown there that the leading term in the expression of variance under H_0 is $\frac{n^3}{6(k+1)}$. Hence using the central limit theorem, it is seen that,

$$n^{-\frac{3}{2}}[W_{RSS}^+ - \frac{n(n+1)}{4}] \xrightarrow{d} N(0, \frac{1}{6(k+1)}) \quad \text{as} \quad n \to \infty. \qquad (5.5.2)$$

Thus the test, for moderately large n, can be carried out based on the above normalized version of the statistic by using the critical points from the standard Normal distribution.

It should be noted that the gain in the efficiency of the RSS procedure over the SRS procedure is of the order $\frac{k+1}{2}$ as that is the ratio of the two asymptotic variances.

5.6   Tests for Randomness

The assumption regarding the data $X_1, X_2, \ldots, X_n$ collected from a probability distribution $F(x)$ is that it is a random sample, which means that these are identically distributed and independent random variables. The test procedures in the earlier sections of this chapter do not question any part of this assumption. Now we consider testing the hypothesis of randomness. To simplify the situation, let us assume that the population consists of only two types of symbols, say $\{H, T\}$ or $\{M, F\}$, etc. Consider the following three sequences of outcomes of size 10 consisting of 5 symbols of each type.

(i) $\{H \ T \ H \ T \ H \ T \ H \ T \ H \ T\}$,

(ii) $\{H\ H\ H\ H\ H\ T\ T\ T\ T\ T\}$,

(iii) $\{H\ T\ T\ H\ T\ H\ H\ H\ T\ T\}$.

The first and the second sequences do not appear to be random as in (i) the symbols H and T alternate and in (ii) the symbols H precede the $T's$. Whereas the third outcome appears to be random more readily. The sequences of two types of symbols may be generated by a random sample from a distribution. Record $+$ for those greater than the sample median (or some other landmark) and $-$ for those below the sample median. In fact, in more involved circumstances one may have three or more types of symbols as well, for example, the roll of a dice gives 6 possible outcomes.

A Run is defined to be a sequence of symbols of the same type preceeded and succeeded by symbols of another type or no symbol at all.

Since lack of randomness is indicated by some pattern in the arrangement of the two types of symbols, one could count R the number of runs of like symbols in the sequence to test the null hypothesis that there is randomness. Let there be n symbols corresponding to the n observations in the random sample and let n_1 and n_2 be the number of symbols of Type I and Type II, respectively. Let

$$R_1 = \text{ be the number of runs of Type I}$$

$$R_2 = \text{ be the number of runs of Type II .}$$

Then, R, the number of runs is given by

$$R = R_1 + R_2.$$

In the 3 sequences given above, the realizations of (R_1, R_2) are $(5,5), (1,1)$ and $(3,3)$, respectively. R takes the values $10, 2$ and 6, respectively. It should be noted that, in general, $r_2 = r_1 - 1, r_1, r_1 + 1$, only.

First we take an approach conditional on a given number of symbols of the two types, viz, n_1 and n_2. Let $n = n_1 + n_2$. We assume that all the $\binom{n}{n_1}$ possible arrangements of the two types of symbols are equally likely to occur under the null hypothesis of randomness. So to find the probability distribution of (R_1, R_2) the number of runs of the two types, we must count the number of arrangements that lead to the various values of (R_1, R_2).

The number of distinguishable ways of distributing n items into r distinct groups is the same as the total number of ways of putting up $r - 1$ barriers in one of the $n - 1$ available distinct places for such barriers. This can be achieved by choosing $(r - 1)$ of $(n - 1)$ places without repetition, which can be done in $\binom{n-1}{r-1}$ ways. Hence one can arrange n_1 symbols of the first kind and n_2 symbols of the second kind in r_1 and r_2 distinct runs in

$\binom{n_1-1}{r_1-1}\binom{n_2-1}{r_2-1}$ ways. However, r_1 can either be $r_2 - 1, r_2, r_2 + 1$. If it is r_2 then one can do so in twice as many ways starting with either of the two types. Hence,

$$P_{H_0}[R_1 = r_1, R_2 = r_2] = c\,\frac{\binom{n_1-1}{r_1-1}\binom{n_2-1}{r_2-1}}{\binom{n}{n_1}},$$

where,

$$c = 2 \text{ if } r_1 = r_2,$$
$$= 1 \text{ if } r_1 = r_2 - 1 \text{ or } r_2 + 1.$$

The above joint distribution leads to the distribution of $R = R_1 + R_2$, the total number of runs, by summing over the probabilities of the relevant sets. Then for $r = 2, 3, \ldots, n$ we have

$$P_{H_0}[R = r] = \begin{cases} 2\dfrac{\binom{n_1-1}{\frac{r}{2}-1}\binom{n_2-1}{\frac{r}{2}-1}}{\binom{n}{n_1}} & \text{if } r \text{ is even,} \\[4mm] \dfrac{\binom{n_1-1}{\frac{r-1}{2}}\binom{n_2-1}{\frac{r-3}{2}}+\binom{n_1-1}{\frac{r-3}{2}}\binom{n_2-1}{\frac{r-1}{2}}}{\binom{n}{n_1}} & \text{if } r \text{ is odd.} \end{cases}$$

The exact distribution of R given above may be used to calculate $E_{H_0}(R)$ and $Var_{H_0}(R)$. But this is tedious. A simpler way for the same is discussed below.

Let us impose the sequence $Y_1, Y_2, \ldots, Y_n$ on the sequence of two types of symbols as follows

$$Y_1 = 1,$$
$$Y_i = \begin{cases} 1, & \text{if the } ith \text{ element is different from the } (i-1)th \text{ element} \\ 0, & \text{if they are identical,} \end{cases}$$

for $i = 2, 3, \ldots, n$. Then, it is obvious that

$$R = \sum_{i=1}^{n} Y_i = 1 + \sum_{i=2}^{n} Y_i.$$

Then, marginally, Y_i is a Bernoulli random variable with

$$E(Y_i) = P(Y_i = 1) = \frac{n_1 n_2}{\binom{n}{2}}, \quad i = 2, 3, \ldots, n.$$

This quickly gives

$$E_{H_0}(R) = 1 + \frac{(n-1)n_1 n_2}{\binom{n}{2}} = 1 + \frac{2n_1 n_2}{n}.$$

Since, $Y_1, Y_2, \ldots, Y_n$ are not independent, the calculation of $Var_{H_0}(R)$ is a little more complicated. But it can be seen that

$$Var_{H_0}(R) = \frac{2n_1n_2(2n_1n_2 - n_1 - n_2)}{n^2(n-1)}.$$

If we let $\lambda = lim_{n \to \infty} \frac{n_1}{n}$, then [Wald and Wolfowitz (1940)] have proved that under H_0,

$$R^* = \frac{R - 2n\lambda(1-\lambda)}{2\lambda(1-\lambda)\sqrt{n}} \xrightarrow{d} N(0,1) \text{ as } n \to \infty.$$

Thus, the test of H_0 would be to reject if $R^* > z_{1-\frac{\alpha}{2}}$ or $R^* < z_{\frac{\alpha}{2}}$ where $z_{(.)}$ denotes the quantile of the specified order of the standard Normal distribution, provided n is large.

The probability of rejection under the alternative of lack of randomness tends to 1, thus showing that the test based on R is consistent.

Comments:
As alluded above, the two types of objects to be arranged in a sequence can arise in many ways indicated below:

(i) outcomes from a sample space with only two points,
(ii) if X is a random variable with median 0, then positive and negative observations,
(iii) if $X_1, X_2, \ldots, X_{n_1}$ and $Y_1, Y_2, \ldots, Y_{n_2}$ are random samples from distributions $F(x)$ and $G(x)$ respectively, then rank all of these in increasing order. The $X's$ and $Y's$ in this order form the runs of the two types of objects.

Example 5.3: Let following be the sequence of genders of people waiting in a queue to purchase movie tickets:

$$MFMFMFFFMFMFMFMMMMFMFFMFF.$$

Here $n = 25, n_1 = 12, n_2 = 13, R = 18, R^* = 1.852$. The value of R^* is greater than the 10% two-sided critical value or one sided 5% critical value of 1.65 from the standard Normal distribution. Hence we can conclude that there is some mild evidence against the null hypothesis that males and females take their place at random in the queue.

In R, the function runs.test from the tseries package can be used to implement this test for randomness.

In this section we have discussed the test based on R, the total number of runs. One may also base tests on (i) the length of the longest run and (ii) the runs up and down, if the sequence consists of real valued observations, or similar information in case the sequence consists of ordinal values. We do not discuss these statistics here.

After the development of the distribution theory of the run statistics under H_0 by Wald and Wolfowitz in the forties of the last century, there was considerable work on investigating power properties of these tests for certain alternatives describing dependence such as Markov dependence by [David (1947)] and [Bateman (1948)]. We refer the interested readers to these papers. [Wald and Wolfowitz (1940)] and [Blum *et al.* (1957)] have established the consistency of the test for the alternative hypothesis $F(x) \neq G(x)$ for some x, where F and G also satisfy a specific condition (called condition **t**).

5.7   Nonparametric Confidence Interval for the Median of a Symmetric Distribution Based on the Signed Rank Statistic

Let the data be $Y_1, Y_2, \ldots, Y_n$. To test the H_0 : median $= 0$ against H_1 : median $\neq 0$ we use the signed rank statistic W^+ given in (5.3.1). The critical region is $|W^+| > w_\alpha$, where w_α is the appropriate critical point. The statistic W^+ can also be developed as below.

Let $\frac{X_i + X_j}{2}$, $i \leq j$ be the $m = \frac{n(n+1)}{2}$ pairwise averages. These are called Walsh averages. The median of these m averages is a point estimate of the population median (or the centre of symmetry). By subtracting this median from each of the observation makes the value of W^+ as close to its null expectation $\frac{n(n+1)}{4}$ as possible. Let

$$u_\alpha = \frac{n(n+1)}{2} + 1 - w_{\frac{\alpha}{2}},$$

where $w_{\frac{\alpha}{2}}$ is the upper $\frac{\alpha}{2}$th quantile of the null distribution of W^+.

Now consider the points θ_L and θ_U from the ranked Walsh averages where θ_L is the $u_\alpha th$ ordered Walsh average and θ_U is the $(m + 1 - u_\alpha)th$ ordered Walsh average. Then if θ is the centre of symmetry (median) of the distribution, then

$$1 - \alpha = P(\theta_L < \theta < \theta_U) \tag{5.7.1}$$

for all θ and for all continuous distributions.

The null distribution of W^+ and the program for the computations required for the test statistic, the point estimate and the confidence interval are available in the R functions wilcox.test and psignrank in the package stats.

5.8 Exercises

(1) Suppose the data are:

$$0.848, \ 0.413, \ 1.058, \ 0.858, \ 0.350, \ 1.582, \ 0.372,$$

$$0.566, \ 1.097, \ 0.939, \ 0.761, \ 0.864, \ 0.712, \ 0.264,$$

$$0.138, \ 1.124, \ 0.383, \ 1.194, \ 2.072, \ 0.690.$$

Use the sign test and the Wilcoxon signed rank test to test the null hypothesis $H_0 : median = 1$ versus the alternative $H_1 : median < 1$. (The data are observations from the Weibull distribution with shape parameter 2 and the scale parameter 1. Thus the true median is 0.833.)

(2) Students in a class were divided in two batches of 10. One batch was given a quiz in the morning before their lectures and the same quiz was given to the other batch in the afternoon after their lectures. The quiz scores out of 10 are given below.

Morning batch scores					Afternoon batch scores				
6,	5,	8.5,	7,	6,	2.5,	10,	10,	4.5,	5.5,
9,	5,	5,	10,	8.5	7,	6,	8,	9.5,	6.5.

Use the run test to determine whether the time of the quiz has an effect on the scores.

(3) Show that the statistic W^+ can be written as a linear combination of two U-statistics.

(4) Prove that $W^+ = \sum_{i=1}^{n} \Psi_i R_i$ and $\sum_{i=1}^{n}(1 - \Psi_i)R_i$ are identically distributed under the null hypothesis.

(5) Prove that the one sided test based on S^- is unbiased.

Chapter 6

OPTIMAL NONPARAMETRIC TESTS

6.1 Introduction

Let us say that the data $X_1, X_2, \ldots, X_n$ is a random sample or a collection of independent and identically distributed random variables from the continuous distribution function F. This may be stated as the following null hypothesis

H_0 : $X_1, X_2, \ldots, X_n$ are i.i.d. random variables with a continuous

c.d.f $F(x)$ and p.d.f $f(x)$. (6.1.1)

In Chapter 2 we have introduced order statistics $X_{(1)}, X_{(2)}, \ldots, X_{(n)}$ of a random sample $X_1, X_2, \ldots, X_n$. Let X_i be the $r_i th$ order statistic, that is, the ith observation happens to be the $r_i th$ ordered observation. We denote the event

$$\{X_i \equiv X_{(r_i)}\} \text{ as } \{R_i = r_i\},$$

that is, R_i being called the (random) rank of the ith observation and r_i is its realization in the random sample.

The vector $\underline{R} = (R_1, R_2, \ldots, R_n)'$ is called the rank vector of the random sample. It takes $n!$ values $\underline{r}$ which are the $n!$ permutations of the vector $\{1, 2, \ldots, n\}$. Let $\underline{X}_{(.)} = (X_{(1)}, X_{(2)}, \ldots, X_{(n)})$ be the vector of order statistics. Then, $\underline{X}_{(.)}$, the vector of the order statistics and $\underline{R}$, the vector of ranks are sufficient to rebuild the sample.

Further, if $X_{(i)}$, the ith order statistic is actually X_{t_i} the t_i^{th} observation in the random sample, then we define t_i to be the antirank of the order statistic $X_{(i)}$. The vector $\underline{t} = (t_1, t_2, \ldots, t_n)'$ and the vector of ranks $\underline{r} = (r_1, r_2, \ldots, r_n)'$ are in 1:1 correspondence. Hence the probability distribution of $\underline{R}$ and $\underline{T}$ will provide the same probabilities for $n!$ sets in their sample spaces, although numbered differently.

Under H_0

$$P(\underline{R} = \underline{r}) = \int \cdots \int_{-\infty < x_{t_1} < \ldots < x_{t_n} < \infty} \prod_{i=1}^{n} f(x_i) dx_i,$$

$$= \frac{1}{n!},$$

by applying the probability integral transformation $Y = F(X)$ and integrating over all the variables successively.

Thus, we have seen that all the $n!$ rank sets $\{\underline{R} = \underline{r}\}$ where $\underline{r}$ is any one of the vectors generated by the permutation of $\{1, 2, \ldots, n\}$, have probability $\frac{1}{n!}$. A union of k such sets where $\alpha \simeq \frac{k}{n!}$ will provide a critical region with size α approximately. The question about which set should form this union will be answered by the probability of these sets under the alternative to the above H_0 that we have in mind. Under an alternative hypothesis, that is, when the n random variables do not have the same distribution (without giving up independence at the moment), the probabilities of $n!$ rank sets are expected to be unequal. We then choose k of these sets to form the critical region which have the largest probabilities giving us the Most Powerful Rank test. Of course there are two major difficulties here. Firstly, the alternative hypothesis may not be known so precisely. Secondly, even if known, it would be a very difficult and time consuming task to calculate the $n!$ probabilities and then sort them in a decreasing order. However, in next section we discuss how to derive such tests.

6.2   Most Powerful Rank Tests Against a Simple Alternative

Consider the following simple alternative hypothesis

H_1 : $X_1, X_2, \ldots, X_n$ are independent random variables with X_i

      having a continuous c.d.f. F_i and p.d.f. f_i, respectively.

Then, one may write the probability of a typical rank set

$$P_{H_1}(\underline{R} = \underline{r}) = P_{H_1}[R_1 = r_1, R_2 = r_2, \ldots, R_n = r_n]$$

$$= \int \cdots \int \prod_{i=1}^{n} f_i(x_i) dx_i, \tag{6.2.1}$$

where the integration is over the region $-\infty < x_{i_1} < x_{i_2} < \ldots < x_{i_n} < \infty$ such that X_i has rank r_i, that is, x_i occupies the r_i^{th} place in the above order, $i = 1, 2, \ldots, n$.

The integral may be rewritten as

$$P_{H_1}(\underline{R} = \underline{r}) = \frac{n!}{n!} \int \cdots \int_{\underline{r}} \frac{\prod_{i=1}^n f_{r_i}(x_{r_i})}{\prod_{i=1}^n f(x_i)} \prod_{i=1}^n f(x_i)dx_i,$$

$$= \frac{1}{n!} E[\prod_{i=1}^n \frac{f_{r_i}(X_{(r_i)})}{f(X_{(i)})}], \qquad (6.2.2)$$

where the expectation E is with respect to the joint distribution of the order statistics $X_{(1)}, X_{(2)}, \ldots, X_{(n)}$ of a random sample from $F(x)$. This is Hoeffding's formula, [Hoeffding (1951)] which may sometimes be useful.

The MP rank test for H_0 against this H_1 is then:

$$\text{Reject } H_0 \text{ whenever } P_{H_1}(\underline{R} = \underline{r}) \geq c_\alpha, \qquad (6.2.3)$$

where c_α is the critical point such that $k \simeq \alpha n!$ of the values of $P_{H_1}(\underline{R} = \underline{r})$ are larger than c_α and the remainng $(n! - k)$ values are smaller, thus simultaneously ensuring that the size of the test is α and the power is maximized. This formulation may be regarded as the marginal version of the Neyman-Pearson lemma for rank tests. It brings out the dependence of the MP test on the (integrated) likelihood ratio of the order statistics. All the same, one must acknowledge that barring exceptional circumstances, such precise knowledge of the distribution under the alternative is not available.

In the parametric setup the probability distributions are within a family and characterized by the value of a real or a vector parameter. There is a popular nonparametric description of departure from the null hypothesis which may adequately describe the experimental situation. The so called Lehmann alternatives first introduced in (1953) say that

$$H_1 : X_i \sim F^{\Delta_i},$$

where $\Delta_1, \Delta_2, \ldots, \Delta_n$ are positive constants, but not all equal. Similarly, the Cox proportional hazards model [Cox (1972)] specified the alternative

$$H_1 : X_i \sim 1 - (1 - F)^{\Delta_i},$$

using one of these in the above formula makes $P_{H_1}(\underline{R} = \underline{r})$ free of basic distribution F and depend only on the constants $\Delta_1, \Delta_2, \ldots, \Delta_n$. Thus these alternatives are called semiparametric. Physically, if Δ_i is a positive integer, then F^{Δ_i} and $1 - (1 - F)^{\Delta_i}$ represent the distributions of the maximum and the minimum of Δ_i i.i.d. random variables with common distribution function F. Besides the second representation describes the proportional hazards model propogated by Cox. Probabilistically these provide families of stochastically monotonically increasing (or decreasing) distributions as Δ_i's change. The probabilities of the rank orders will not depend on F but only on the Δ_i's.

Let the order statistics of the random sample $X_1, X_2, \ldots, X_n$ be $X_{i_1} < X_{i_2} < \ldots < X_{i_n}$ where $(i_1, i_2, \ldots, i_n)$ is a permutation of $(1, 2, \ldots, n)$. Then, the vector $(i_1, i_2, \ldots, i_n)$ is called the vector of antiranks of the random sample, meaning that X_{i_1} has rank 1, X_{i_2} has rank 2, $\ldots$, X_{i_n} has rank n. There is a 1-1 correspondence between the rank vectors and the antirank vectors. [Savage (1956)] has obtained the probabilities of these $n!$ vectors of antiranks. Since there is 1:1 connection between the vector of ranks and antiranks, these are also the probabilities of the rank vectors.

$$P_{H_1}(\underline{R} = \underline{r}) = P_{H_1}(X_{i_1} < X_{i_2} < \ldots < X_{i_n})$$
$$= \frac{\prod_{i=1}^{n} \Delta_i}{\prod_{k=1}^{n} (\sum_{j=1}^{k} \Delta_{i_j})}, \qquad (6.2.4)$$

where $(i_1, i_2, \ldots, i_n)$ is the vector of antiranks corresponding to the rank vector $\underline{r}$.

Apart from the Lehmann alternatives it is almost impossible to obtain the probabilities of the rank orders, thus preventing us from deriving useful most powerful rank tests. Also it is only seldom that we will know the exact probability distribution for which we need the rank order probabilities.

6.3   Locally Most Powerful (LMP) Rank Tests

Let the random sample $X_1, X_2, \ldots, X_n$ have a joint pdf $f(x_1, x_2, \ldots, x_n, \Delta)$, where Δ is a real valued parameter. It could be the familiar location/scale or some other parameter. What we require is that it should distinguish between the null and the alternative hypotheses in the following manner

$$H_0 : \Delta = 0 \quad \text{and} \quad H_1 : \Delta > 0.$$

We assume that the null hypothesis H_0, that is, $\Delta = 0$ implies that $X_1, X_2, \ldots, X_n$ are i.i.d. random variables from a continuous distribution and the alternative hypothesis H_1 ($\Delta > 0$) excludes this case. For example, one may have

$$f(x_1, x_2, \ldots, x_n, \Delta) = \prod_{i=1}^{n} f(x_i, c_i \Delta), \quad c_i \neq 0 \ \forall \ i.$$

Thus, $\Delta = 0$ leads to $\prod_{i=1}^{n} f(x_i, 0)$, the joint distribution of i.i.d. random variables and $\Delta \neq 0$ gives independent but not identically distributed random variables.

As mentioned earlier, it would be difficult to obtain the probability for value of $\Delta \neq 0$. In this situation we note that, by Taylor expansion

$$P_\Delta(\underline{R} = \underline{r}) = \frac{1}{n!} + \Delta \frac{\partial}{\partial \Delta} P_\Delta(\underline{R} = \underline{r})|_{\Delta=0} + o(\Delta)$$

$$= \frac{1}{n!} + \Delta T(\underline{r}) + o(\Delta), \qquad (6.3.1)$$

where the statistic $T(\underline{r}) = \frac{\partial}{\partial \Delta} P_\Delta(\underline{R} = \underline{r})|_{\Delta=0} \neq 0$. So for small values of $\Delta \; (> 0)$, that is, for alternatives near to the null hypothesis, one can say approximately that

$$P_\Delta(\underline{R} = \underline{r}) > c_\alpha$$

is the same as

$$T(\underline{r}) > k_\alpha,$$

for appropriate k_α. $T(\underline{r})$ is essentially the slope of the power as a function of Δ at $\Delta = 0$. Hence the test would be size α test which maximizes the slope of the power function at $\Delta = 0$. This is the generally accepted concept of a locally most powerful test; a LMP rank test in our context.

We shall see that $T(\underline{r})$ is very often a manageable function of $\underline{r}$ and simple tests emerge of this approach. Using (6.2.2) we can write

$$T(\underline{r}) = \frac{\partial}{\partial \Delta} E[\prod_{i=1}^{n} \frac{f(X_{(r_i)}, \Delta)}{f(X_{(i)})}], \qquad (6.3.2)$$

evaluated at $\Delta = 0$. We see that this is a linear rank statistic of the form

$$T(\underline{r}) = \sum_{i=1}^{n} g(r_i), \qquad (6.3.3)$$

where $g(r_i)$ is a function of r_i, the rank of the ith observation alone.

6.4 Rank Statistics and Its Null Distribution

It is clear that $\underline{R}$, the vector of rank statistics, provides a very natural base for the construction of nonparametric methodology. The probability integral transformation $Y = F(X)$, where F is the c.d.f. of X leads to the easy calculation of probabilities of the rank set, that is, the probability distribution of $\underline{R}$ under the null hypothesis of $X_1, X_2, \ldots, X_n$ being i.i.d. continous random variables.

Later we will see that many of the important rank test statistics turn out to be of the form

$$T(\underline{r}) = \sum_{i=1}^{n} c_i a(R_i). \qquad (6.4.1)$$

This notation is originally from [Hájek and Šidák (1967)]. The $c_i's$ are called regression constants and $a(R_i)'s$ are called scores based on rank R_i, of X_i $i = 1, 2, \ldots, n$ in the random sample.

Theorem 6.1. Under H_0 of section 6.1:

$$E(S) = n\bar{c}\bar{a}$$

$$Var(S) = \frac{1}{n-1}[\sum_{i=1}^{n}(c_i - \bar{c})^2 \sum_{i=1}^{n}(a_i - \bar{a})^2], \qquad (6.4.2)$$

where $\bar{a} = \frac{1}{n}a_i$, $\bar{c} = \frac{1}{n}c_i$.

Proof: The proof of the expectation follows from the fact that

$$E(S) = \sum_{i=1}^{n} c_i E(a(R_i))$$

and $a(R_i)$ has the discrete uniform distribution over $(a_1, a_2, \ldots, a_n)$, due to the fact that

$$P(R_i = k) = \frac{\{\text{number of permutations in which } R_i = k\}}{\{\text{Total number of permutations}\}}$$

$$= \frac{(n-1)!}{n!} = \frac{1}{n}.$$

Then

$$V(S) = \sum_{i=1}^{n} c_i^2 V(a(R_i)) + 2 \sum_{1 \le i < j \le n} c_i c_j Cov(a(R_i), a(R_j)). \qquad (6.4.3)$$

Due to discrete uniform distribution of R_j over $\{1, 2, \ldots, n\}$ we can obtain

$$Var(a(R_i)) = \frac{1}{n} \sum_{i=1}^{n}(a(i) - \bar{a})^2,$$

and

$$Cov(a(R_i), a(R_j))) = -\frac{1}{n(n-1)} \sum_{i=1}^{n}(a(i) - \bar{a})^2.$$

Substituting these in (6.4.3) and a little rearrangement gives the variance formula of the theorem.

Although the terms in the sum of the linear rank statistic are not independent, the standardized version of the statistic has asymptotically a standard normal distribution under any one of the following conditions. The conditions are on both the regression constants and the scores. They are

(i) Noethers's condition:

$$\frac{\sum_{i=1}^{n}(c_{ni} - \bar{c}_n)^2}{\max_{1 \leq i \leq n}(c_{ni} - \bar{c}_n)^2} \to \infty \text{ as } n \to \infty.$$

Here the constants are dependent on n, the number of observations and the condition essentially says that no individual constant should be too far away (dominate) from the mean.

(ii) Hajek's condition:

Let us assume that the scores $a(i)'s$ are generated by function $\Phi(u)$, $0 < u < 1$, that is,

$$a_n(i) = \Phi(\frac{1}{n+1}), \quad i = 1, 2, \ldots, n.$$

Then, the condition says that the $\Phi(u)$ is a square integrable score function, that is,

$$0 < \int_0^1 (\Phi(u) - \bar{\Phi})^2 du < \infty,$$

where $\bar{\Phi} = \int_0^1 \Phi(u) dy$.

Theorem 6.2. Let $T_n(\underline{R}) = \sum_{i=1}^{n} c_{ni} a_n(R_i)$ be a linear rank statistic where the constant $c's$ and the score $a's$ satisfy one of the above two conditions. Then

$$T_n^*(\underline{R}) = \frac{T_n(\underline{R}) - ET_n(\underline{R})}{\sqrt{V(T_n(\underline{R}))}} \xrightarrow{d} N(0, 1) \text{ as } n \to \infty.$$

Proof: The proof depends on an intermediate lemma which proves the asymptotic normality of the standardized version of the statistic $\sum_{i=1}^{n} c_{ni} a_{ni}(U_i)$ where $U_1, U_2, \ldots, U_n$ are i.i.d. uniform (0,1) random variables. Then the next stage is to prove that $T_n^*(\underline{R})$ and $T_n^*(\underline{U})$ have asymptotically the same distribution. For details see [Hájek and Šidák (1967)], [Randles and Wolfe (1979)], etc.

As indicated at various places above, this proof depends upon the discrete uniform distribution of $\underline{R}$ which holds under the null hypothesis of this chapter. It is also noted that the discrete uniform distribution of $\underline{R}$ over the $n!$ permutations of $\{1, 2, \ldots, n\}$ holds under the more general continuous exchangeable joint distribution for $(X_1, X_2, \ldots, X_n)$.

We shall point out in subsequent chapters that the popular Wilcoxon-Mann-Whitney, Fisher-Yates-Terry-Hoeffding normal scores test and many other tests use constants and scores corresponding to the LMP rank tests in certain cases and which also satisfy the conditions (i) and (ii) and have asymptotically normal distributions under the null hypothesis.

6.5   Exercises

1. Let $X_1, X_2, \ldots, X_n$ be i.i.d. random variables from a continuous distribution $F(x)$ and $Y_1, Y_2, \ldots, Y_m$ from $F(x - \theta)$, $\theta > 0$. Find the scores for the Locally Most Powerful Rank test when

(i) X has standard Normal distribution,
(ii) X has standard logistic distribution.

2. Let $X_1, X_2, \ldots, X_n$ be i.i.d. random variables from a continuous distribution $F(x\theta)$, $\theta > 1$ and $Y_1, Y_2, \ldots, Y_m$ from $F(x)$. Find the scores for the Locally Most Powerful Rank test when

(i) X has standard Normal distribution,
(ii) X has standard exponential distribution.

Chapter 7

THE TWO SAMPLE PROBLEM

7.1 Introduction

Historically, after the chi-square goodness of fit test and Spearman's rank correlation discussed in Chapters 4 and 9, respectively, the two sample Wilcoxon-Mann-Whitney test [Wilcoxon (1945)], [Mann and Whitney (1947)] was the major contribution to nonparametric methods.

Let us describe the two sample problem first. The data consists of $X_1, X_2, \ldots, X_{n_1}$ and $Y_1, Y_2, \ldots, Y_{n_2}$, two independent random samples from probability distributions $F(x)$ and $G(x)$, respectively. The null hypothesis to be tested is

$$H_0 : F(x) = G(x) \ \forall \ x. \tag{7.1.1}$$

Thus, under the null hypothesis, all the $n = n_1 + n_2$ observations come from the common distribution, say, $F(x)$ and form a random sample of size n from it. We further assume that $F(x)$ and $G(x)$ are continuous cumulative distribution functions. This is a direct extension of the field of applicability of the two sample t-test. The t-test is used for comparing the Normal probability distributions with different means and with the same variance. Now we do not restrict to the normal distribution family.

Under the null hypothesis all the n observations constitute a random sample, the theory of Optimal Rank Tests developed in Chapter 6 applies readily here. Let us take up tests appropriate for the various two sample alternative hypothesis - location and scale one by one. Two sample test for goodness of fit is also discussed.

7.2   The Location Problem

Recall that the two sample t-test is meant to detect the difference in means (locations) of two normal distributions. Without assuming normality let us set up this problem as

$$H_0 : \ F(x) = G(x) \ \forall \ x,$$

vs

$$H_1 : \ F(x) = G(x - \Delta) \ \forall \ x, \ \ \Delta \geq 0. \qquad (7.2.1)$$

If the densities corresponding to $F(x)$ and $G(x)$ exist then the following Figure 7.1 would indicate the situation under H_1.

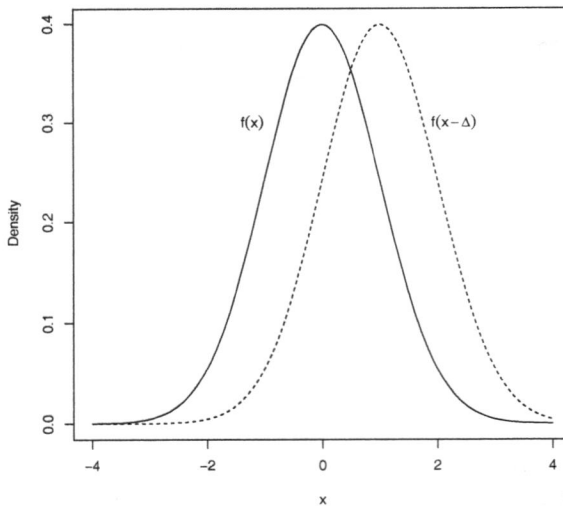

Fig. 7.1   Densities with Location Shift

It is indicated that the densities differ only in their location parameters and will be identical if $\Delta = 0$. We use expression (6.3.2) to derive the statistic on which the locally most powerful rank test would be used. The approach of LMP rank tests allows us to derive tests which are appropriate in the two sample location case for certain known families.

We see that for a rank order $\underline{r} = (r_1, r_2, \ldots, r_n)$,

$$P_\Delta(\underline{R} = \underline{r}) = n_1! n_2! \int \cdots \int_{\underline{R} = \underline{r}} f(w_i - \Delta c_i) dw_i, \qquad (7.2.2)$$

where $c_i = 0$ for the first sample and $c_i = 1$ for the second sample. Then, under regularity conditions which allow interchange of the integration and the limit (as $\Delta \to 0$) operations, we obtain

$$T(\underline{R}) = \sum_{i=1}^{n} c_i E\left(\frac{-f'(w_i)}{f(w_i)}\right)$$

$$= \sum_{i=n_1+1}^{n_1+n_2} E\left[\frac{-f'(w_{(R_i)})}{f(w_{(R_i)})}\right], \qquad (7.2.3)$$

as given in (6.3.2) where $R_{n_1+1}, R_{n_1+2}, \ldots, R_{n_1+n_2}$ are the ranks of the second sample in the combined order and $w_{(1)}, w_{(2)}, \ldots, w_{(n_1+n_2)}$ are the order statistics corresponding to the combined random sample. The expectations of the ratio, called scores, can be easily calculated for many distributions. For details of these derivations refer to [Hájek and Šidák (1967)] or [Govindarajulu (2007)].

Two common and classic nonparametric tests are seen to be LMP rank tests through this approach.

7.2.1 *The Wilcoxon Rank Sum Test*

Suppose that we wish to derive the LMP rank test for the logistic distribution. Then

$$f(x) = (1 + e^{-x})^{-2}.$$

This can also be expressed as

$$f(x) = F(x)(1 - F(x)).$$

This enables us to explicitly calculate

$$\frac{-f'(x)}{f(x)} = (2F(x) - 1).$$

Then $T(\underline{R})$ is a linear function of

$$W = \sum_{j=1}^{n_2} R_{n_1+j}, \qquad (7.2.4)$$

the rank sum of the second random sample. Besides being the LMP rank test for the logistic distribution, the Wilcoxon test has a very high asymptotic relative efficiency (ARE) of 0.95 against the two sample t-test for the

Normal distribution. It is obvious that it is very simple to apply. Its mean and variance under the null hypothesis can be obtained from Theorem 6.1 as

$$E_{H_0}(W) = \frac{n_2(n+1)}{2},$$

$$Var_{H_0}(W) = \frac{n_1 n_2(n+1)}{12}. \tag{7.2.5}$$

If there are ties in the observations, say in k groups with $t_1, t_2, \ldots, t_k$ observations in each group, then assigning the midrank to each of the observations in the tied group reduces the variance to

$$Var_{H_0,T}(W) = \frac{n_1 n_2(n+1)}{12} - \frac{n_1 n_2 \sum_{i=1}^{k} t_i(t_i^2 - 1)}{12n(n-1)}.$$

Further Theorem 6.2 assures that under H_0,

$$W^* = \frac{W - E_{H_0}(W)}{\sqrt{Var_{H_0}(W)}} \xrightarrow{d} N(0,1) \text{ as } n \to \infty.$$

This allows the use of the standard normal tables for the choice of critical points when sample size is large

7.2.2 The Mann-Whitney Test

The Wilcoxon rank sum statistic W given in (7.2.4) is linearly related to the Mann-Whitney statistic U defined as

$$U = \sum_{i=1}^{n_1} \sum_{j=1}^{n_2} I(X_i \leq Y_j), \tag{7.2.6}$$

where I is the indicator function of the event.

It is readily seen that

$$U = \sum_{j=1}^{n_2} (R_j - m_j), \tag{7.2.7}$$

where R_j is the rank of Y_j $(j = 1, 2, \ldots, n_2)$ and m_j is the number of observations from the Y-sample which are less than or equal to Y_j. Hence $R_j - m_j$ is the number of pairs (X_i, Y_j) in which the X observation is smaller than the Y observation, and their sum is the sum of the indicator function of all such events. Therefore,

$$U = W - \sum_{j=1}^{n_2} j = W - \frac{n_2(n_2+1)}{2}. \tag{7.2.8}$$

Hence the formula for the mean and variance of U can be derived from those of W and the asymptotic normality of U too follows.

7.2.3 *Comments*

(i) Although we have described the Wilcoxon-Mann-Whitney test as the LMP rank test for the logistic distribution, it should be understood that it predates the development of LMP rank tests. One of the motivations was to seek tests for the alternative of 'stochastic dominance'. Let

$$H_0 : F(x) = G(x) \ \forall \ x$$

vs

$$H_1 : F(x) \geq G(x) \ \forall \ x,$$

with a strict inequality holding for some x. The random variable Y is said to stochastically dominate the random variable X. We illustrate the alternative hypothesis in Figure 7.2 below.

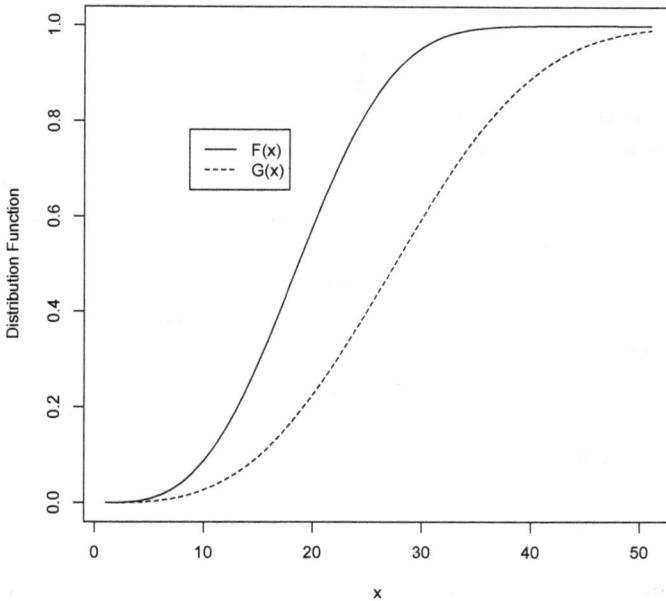

Fig. 7.2   Stochastic Dominance

The alternative hypothesis says that $F(x) \geq G(x) \ \forall \ x$, or the probability that the random variable X is less than or equal to x is larger than the probability that the random variable Y is less than or equal to the same number x, for every x. Or the random variable Y tends to take a larger value with a larger probability than the random variable X. This is a larger class of hypothesis than the location alternative hypothesis. It is seen that

$$
\begin{aligned}
P_{H_1}[X \leq Y] &= \int_{-\infty}^{\infty} F(y)dG(y) \\
&\geq \int_{-\infty}^{\infty} G(y)dG(y) \\
&= \frac{1}{2} \\
&= P_{H_0}[X \leq Y].
\end{aligned}
$$

(ii) We see that

$$
\begin{aligned}
E(\frac{U}{n_1 n_2}) &= E(I(X_i \leq Y_j)) \\
&= P(X \leq Y).
\end{aligned}
$$

Hence U is able to discriminate between the H_0 and the extended alternative H_1 of stochastic dominance. This heuristic observation provides a justification for the use of the Mann-Whitney U test (and equivalently for the Wilcoxon W test) for this testing problem, apart from it being the LMP rank test for the Logistic distribution.

(iii) Essentially, it is being argued that observations coming from a distribution with large mean, or one which leads to larger observations with a large probability will yield observations with larger ranks. Hence large values of the sum of ranks of these observations support the alternative hypothesis. The theory of rank tests of Chapter 6 provides the null distribution of the test statistic which is entirely distribution free for continuous random variables.

7.3   Two Other Tests

In Chapter 6 it was noted that many of the important rank test statistics turn out to be of the form

$$
T(\underline{R}) = \sum_{i=1}^{n} c_i a(R_i), \tag{7.3.1}
$$

where
$$c_i = \begin{cases} 0 & \text{for } i = 1, 2, \ldots, n_1, \\ 1 & \text{for } i = n_1 + 1, n_1 + 2, \ldots, n_1 + n_2. \end{cases}$$

Thus, $T(\underline{R})$ represents the sum of scores corresponding to the second sample observations. In the Wilcoxon test described in section (7.2) the ranks are the scores. Expression (6.3.2) gives the scores for the LMP rank tests for different distributions.

Here we describe two more systems of scores which have been suggested in the literature and are used in practice.

(i) The scores proposed by Fisher-Yates-Terry [Terry (1952)] and [Hoeffding (1951)] are the expected values of the normal order statistics,

$$a(i) = E(V_{(i)}), \tag{7.3.2}$$

where $V_{(i)}$ is the ith order statistic of a random sample of size n from the $N(0, 1)$ distribution. These arise as the appropriate scores for the LMP rank test for the normal distribution.

Thus the statistic is

$$C_1 = \sum_{i=1}^{n} c_i E(V_{(R_i)}), \tag{7.3.3}$$

with $c_i = 0$ for the X observations and 1 for the Y observations. Effectively, it is the sum of the expectations of normal order statistics corresponding to those order statistics in the combined order, which correspond to the second random sample. These expected order statistics, not being scores as simple as the ranks themselves, need to be tabulated for various n, or their computation needs to be included in the software to be used for calculating the statistics C_1 and carrying out the test based on this set.

(ii) The second system of scores is proposed by Van der Waerden and is

$$a(i) = \Phi^{-1}\left(\frac{i}{n+1}\right), \tag{7.3.4}$$

where Φ is the cumulative distribution function corresponding to the standard normal distribution. These scores are the asymptotic expectation of the order statistics used as scores in Fisher-Yates-Terry-Hoeffding test statistics given in (7.3.2). The test statistic is once more

$$V_2 = \sum_{i=1}^{n} c_i \Phi^{-1}\left(\frac{R_i}{n+1}\right),$$

with c_i being indicators of the second sample observations as before. The values of the scores are more easily available than the normal scores suggested by Fisher et al. But the null mean and variance of both these statistics C_1 and V_2 is a bit more complicated to compute than that of the Wilcoxon statistics (7.2.4).

The general theory of rank tests ensures that as $n \to \infty$

$$\frac{C_1 - E_{H_0}(C_1)}{\sqrt{Var_{H_0}(C_1)}} \quad \text{and} \quad \frac{V_2 - E_{H_0}(V_2)}{\sqrt{Var_{H_0}(V_2)}},$$

are asymptotically normally distributed with mean 0 and variance 1. This allows us to use the standard normal distribution tables for obtaining the critical points for the tests. These are expected to be quite accurate for even moderately large sample sizes.

The null variances of the two test statistics are

$$Var_{H_0}(C_1) = \frac{n_1 n_2}{(n_1 + n_2)(n_1 + n_2 - 1)} \sum_{i=1}^{n_1+n_2} [E(X_{(i)})]^2,$$

and

$$Var_{H_0}(V_2) = \frac{n_1 n_2}{(n_1 + n_2)(n_1 + n_2 - 1)} \sum_{i=1}^{n_1+n_2} [\Phi^{-1}(\frac{i}{n_1 + n_2 + 1})]^2.$$

Let

$$lim_{n_1,n_2 \to \infty} \frac{n_1}{n} = \lambda \quad (0 < \lambda < 1).$$

Asymptotically, $Var_{H_0}(n^{\frac{1}{2}} C_1)$ as well as $Var_{H_0}(n^{\frac{1}{2}} V_2)$ are equal to

$$\lambda(1 - \lambda) \int_0^1 [\Phi^{-1}(u)]^2 du = \lambda(1 - \lambda),$$

since $\Phi(x)$ is the standard normal distribution function.

All the above tests are easy to apply, there may be some difficulty in obtaining the values of $E(X_{(i)})$, $i = 1, 2, \ldots, n$. So we demonstrate the use of Van der Waerden statistic.

Example 7.1: Let the data be such that $n_1 = 10$, $n_2 = 5$ and $2, 5, 6, 8$ and 9 are the ranks of the observations from the second random sample. Then we need the value of $\Phi^{-1}(\frac{i}{16})$ for these 5 choices of i to calculate the value of the test statistic and for all $i = 1, 2, \ldots, n$ for its variance. Note that $\Phi^{-1}(\frac{i}{n+1}) = -\Phi^{-1}(1 - \frac{i}{n+1})$ because of the symmetry of the standard normal distribution.

Table 7.1 - Calculation of Van der Waerden statistic

i	$\Phi^{-1}(\frac{i}{16})$	i	$\Phi^{-1}(\frac{i}{16})$
1	-1.534	9	.157
2	-1.150	10	.318
3	-.8871	11	.488
4	-.674	12	.674
5	-.488	13	.887
6	-.318	14	1.150
7	-157	15	1.534
8	0		

$V_2 = -1.799$ and the exact variance, under H_0, is 2.514, giving us the standardized value of $V_2^* = \frac{-1.799}{\sqrt{2.514}} = -1.134$. This value is well within the acceptance region of a two-sided test at 5%, or even at 10% level of significance. Thus there is no evidence to reject the null hypothesis of equality of the two distribution functions.

This test can be implemented by using the function 'posthoc.vanWaerden.test' from the package 'PMCMR' in 'R'.

Comments:

(i) Both the above tests perform very well if the data happens to be actually from the normal distribution. The asymptotic relative efficiency of the Van der Waerden test is 1 w.r.t. the two sample t-test.

(ii) This is remarkable since one of the early objections to nonparametric tests used to be their alleged lack of power. Here we see that these tests are fully efficient compared to the two sample t-test and they retain the probability of Type I error, even when the data does not follow the normal distribution. As a matter of fact the ARE of these Fisher-Yates (1932), [Terry (1952)], Hoeffding (1951) test compared to the t-test is not less than 1 for any distribution at all. The ARE of the Wilcoxon-Mann-Whitney [Wilcoxon (1945)] test compared to the two sample t-test is not less than .864 for any distribution, [Lehmann (1951)] and [Chernoff and Savage (1958)] proved these astounding results which completely refute the allegations of low power.

7.4   Introduction to the Two Sample Dispersion Problem

So far, in the earlier sections of this chapter, we have considered the location problem. The two distributions $F(x)$ and $G(x)$ from which the data has arisen are identical (that is, $H_0 : F(x) = G(x) \quad \forall\, x$) or differ only by their locations, that is, $H_1 : F(x) = G(x - \theta) \quad \forall\, x,\ \theta \neq 0$.

Now we consider the two sample dispersion or scale problem. Again the null hypothesis is

$$H_0 \ : \ F(x) = G(x) \quad \forall\, x$$

against the alternative hypothesis

$$H_1 : F(x) = G(\theta x) \quad \forall\, x,\ \theta > 0,\ \theta \neq 1.$$

$$(7.4.1)$$

The following Figure 7.3 will demonstrate the relative positions of $F(x)$ and $G(x)$ under the scale model.

The above description is somewhat restrictive in the following sense. The two c.d.f.'s have the same value, that is, achieve the same probability at $x = 0$. Besides the procedures given below will be seen to work well when this common quantile is the median, that is, essentially saying that the locations are 0, that is, $F(0) = G(0) = \frac{1}{2}$. A generalization is to consider the problem of testing

$$H_0 : \qquad F(x) = G(x) \ \forall\ x$$

against

$$H_1 : F(x - m) = G(\theta(x - m)) \ \forall\ x,$$

where m, rather than 0, is the common median. So we either assume that the two distributions have median 0, or can be adjusted to have median 0. In the Normal distribution case one would use the F test based on the ratio of two sample variances of the two random samples. Here we suggest two tests based on linear rank statistics and a third one which may be regarded as a generalization of the Mann-Whitney U-statistic approach.

While retaining the restriction that the common median or another quantile is known, one could extend the alternative hypothesis in a non-parametric manner. One may say that the alternative hypothesis is

$$H_1 : F(x) \leq G(x) \quad \forall\, x \leq 0 \text{ and} \qquad F(x) \geq G(x) \quad \forall\, x > 0. \qquad (7.4.2)$$

This alternative is illustrated in Figure 7.4.

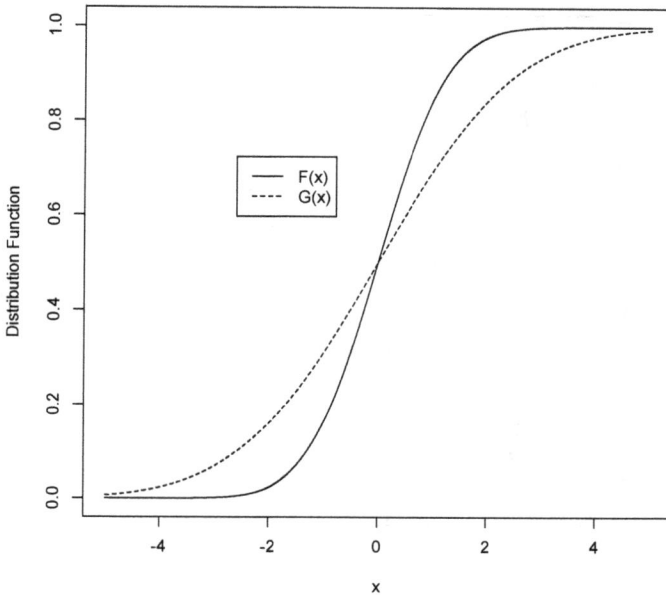

Fig. 7.3   Distributions with Scale Shift

Thus we do not specify any parametric relationship to describe the alternative hypothesis, only specify that under the alternative hypothesis the distribution F gives a larger probability to any interval around 0, than the distribution G.

7.5   The Mood Test

The Mood test statistic fits neatly in the class of linear rank statistics of the type (6.4.1). Rank all the observations from the two samples together from the smallest to the largest, $R_1, R_2, \ldots, R_n$ being the ranks of $X_1, X_2, \ldots, X_{n_1}$ and $Y_1, Y_2, \ldots, Y_{n_2}$ taken together. Define

$$M_n = \sum_{i=1}^{n} c_i a(R_i),$$

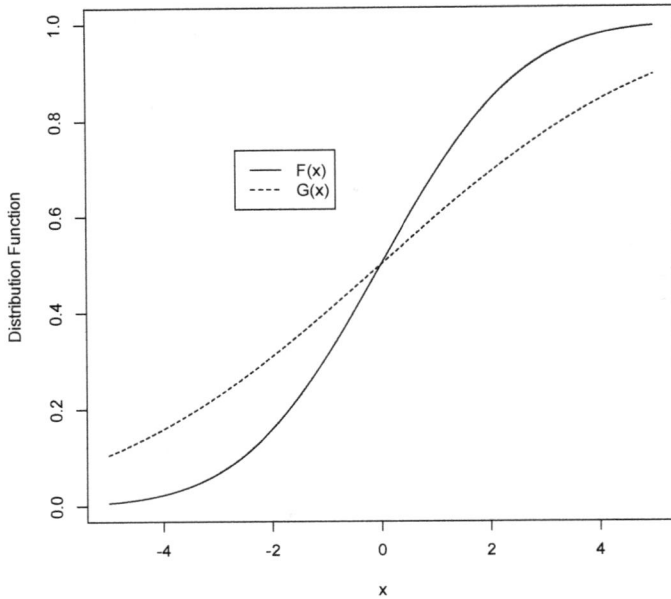

Fig. 7.4   Distributions with one giving larger probability to intervals around '0'

where

$$c_i = 1 \ \text{ for } \ i = n_1 + 1, \ldots, n$$
$$= 0 \ \text{ for } \ i = 1, \ldots, n_1,$$

and

$$a(R_i) = (R_i - \frac{n+1}{2})^2. \tag{7.5.1}$$

Therefore, M_n is the sum of squares of deviations of the ranks of Y observations from $\frac{n+1}{2}$, the average of ranks of all n observations. It is clear that the statistic will be large if the random sample from the second population has a larger number of observations further away from the common quantile 0, compared to the first sample. This should be the case when H_1 is true. Hence the test would be to reject this H_0 if M_n is too large. The critical region should be appropriately modified for the other one sided alternative or for the two-sided alternative.

Using Theorem 6.1 for the statistic M_n with scores given in (7.5.1) and using standard formulas for the sums of positive integers, their squares, cubes and fourth powers, we get the following simple expressions for its mean and variance under the null hypothesis.

$$E(M_n) = \frac{n_2(n^2 - 1)}{12},$$

and

$$Var(M_n) = \frac{n_1 n_2(n + 1)(n^2 - 4)}{180}.$$

The normal approximation holds for the null distribution of

$$M_n^* = \frac{M_n - E_{H_0}(M_n)}{\sqrt{Var_{H_0}(M_n)}},$$

and appropriate one-sided or two-sided critical points from the standard normal distribution may be used to carry out the test when the sample sizes are large.

This test can be carried out by using the function 'mood.test' in the package 'stats' in 'R'.

7.6   The Capon and the Klotz Tests

The Capon test is also based on a linear rank statistic which uses the squares of the expected values of standard Normal order statistics as scores. Thus

$$C_n = \sum_{i=1}^{n} c_i E(V_{(R_i)})^2,$$

where $V_{(i)}$ is the ith order statistic from a random sample of size n from the standard normal distribution. The test based on this statistic is the locally most powerful rank test for the scale alternatives (for the Normal distribution with common median). Later *Klotz* proposed an approximate and asymptotically equivalent system of scores as

$$a(i) = [\Phi^{-1}(\frac{i}{n+1})]^2. \tag{7.6.1}$$

The test statistic is

$$K_n = \sum_{i=1}^{n} c_i \Phi^{-1}(\frac{R_i}{n+1})^2,$$

The asymptotic normality of both the statistics follow from the general limit theorem for linear rank statistics. The means and variances of the two statistics C_n and K_n, under H_0, are

$$E(C_n) = n_2,$$

$$Var(C_n) = \frac{n_1 n_2}{n(n-1)} \sum_{i=1}^{n} (E(V_n^{(i)})^2)^2 - \frac{n_1 n_2}{(n-1)},$$

and

$$E(K_n) = \frac{n_2}{n} \sum_{i=1}^{n} \Phi^{-1}(\frac{i}{n+1})^2,$$

$$Var(K_n) = \frac{n_1 n_2}{n(n-1)} \sum_{i=1}^{n} [\Phi^{-1}(\frac{i}{n+1})]^4 - \frac{n_1}{n_2(n-1)} [E(K_n)]^2.$$

Then, under H_0, both

$$\frac{C_n - E_{H_0}(C_n)}{\sqrt{Var_{H_0}(C_n)}} \quad \text{and} \quad \frac{K_n - E_{H_0}(K_n)}{\sqrt{Var_{H_0}(K_n)}}$$

have asymptotically standard normal distribution. Therefore, standard normal critical points may be used for moderately large sample sizes for these tests.

Comment: The tests considered in this chapter so far are useful when the two distributions have a common median and possibly different scale parameters. If the medians are not equal, then we carry out 'studentization' through subtracting the sample median of the combined random sample. Often, the asymptotic normality of the test statistics still holds.

Example 7.2: Fifty boys were divided into two groups of twenty five boys each. Boys in group I are subjected to exhortations to do well in a test where as those in group II were not subjected to such exhortations. It was felt that variability will be more in the first group than the second. The observed scores are:

The Mood statistic for this data is $M_n = 5936.25$. The null mean and variance are 5206.25 and 442000, respectively. The standardized value is 1.10. This is within the acceptance region for $\alpha = .05$ or even $\alpha = .1$ one tailed test. Hence it can not be concluded that the variability in the first group is more than the variability in the second group.

Any of the other test statistics may also be used in this problem by using the appropriate scores. The scores are given in the table on the next page.

Exhortation					No Exhortation				
139	360	295	360	335	360	49	140	120	162
130	181	91	182	203	131	129	249	38	44
153	360	155	225	71	82	195	47	138	65
124	38	36	203	294	287	54	133	62	220
175	360	360	45	189	131	118	98	131	90

7.7 The Sukhatme-Deshpande-Kusum Tests

At the end of section 7.4 we introduced the following problem. Let $F(x)$ and $G(x)$ be the cumulative distribution functions of X and Y, respectively. Under H_0 they are identical. However, a way of stating that their dispersions are different is to say that upto a point m_0 the distribution function $F(x)$ is below the distribution function $G(x)$ and beyond m_0 the distribution function $G(x)$ is below the distribution function $F(x)$. This may be stated as

$$H_0: \qquad F(x) = G(x) \quad \forall\, x$$

and

$$H_1: F(x) \leq G(x) \quad \forall\, x \leq m_0, \quad F(x) \geq G(x) \quad \forall\, x > m_0.$$

The alternative is illustrated in Figure 7.5. If the point m_0 is such that $F(m_0) = G(m_0) = 0$ or $F(m_0) = G(m_0) = 1$, then it becomes the stochastic dominance problem. However, if $F(m_0) = G(m_0) \neq 0, 1$ then, it is a generalization of the scale problem.

Originally [Sukhatme (1958)] had proposed a test where m_0 was envisaged to be the common median of the two distributions. [Deshpande and Kusum (1984)] proposed an extension which works well where m_0 is the common quantile of any order α ($\alpha \neq 0, 1$). It is based on a U-statistic which is an extension of the Mann-Whitney U-statistic for the stochastic dominance alternative. Let $X_1, X_2, \ldots, X_{n_1}$ and $Y_1, Y_2, \ldots, Y_{n_2}$ be independent random samples from $F(x)$ and $G(x)$, respectively. w.l.o.g. take $m_0 = 0$.

We define a kernel

$$\phi(x_i, y_j) = \begin{cases} 1 & \text{if } 0 \leq x_i < y_j \text{ or } y_j < x_i \leq 0, \\ 0 & \text{if } x_i < 0 < y_j \text{ or } y_j < 0 < x_i \text{ or } x_i = y_j, \\ -1 & \text{if } 0 \leq y_j < x_i \text{ or } x_i < y_j \leq 0. \end{cases}$$

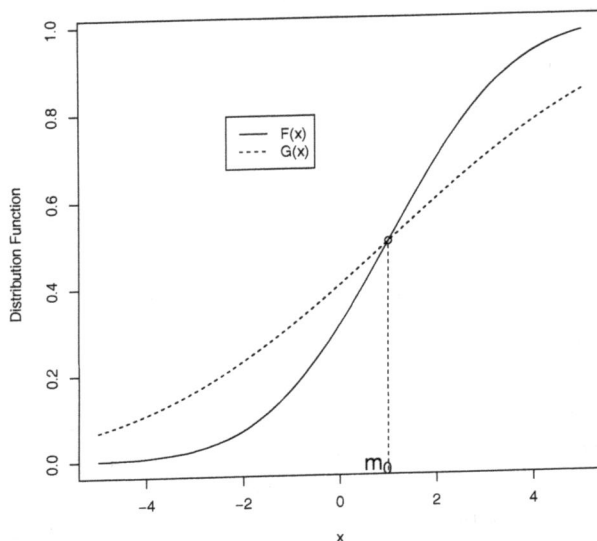

Fig. 7.5   Distributions with unequal dispersions

The U-statistic based on this kernel is

$$T = \frac{1}{n_1 n_2} \sum_{i=1}^{n_1} \sum_{j=1}^{n_2} \phi(x_i, y_j). \qquad (7.7.1)$$

If H_1 is true, we would expect T to take large values. Hence H_0 is to be rejected if T is greater than an appropriate critical value. By exchanging the role of F and G we specify the other one-sided alternative hypothesis. The two sided alternative hypothesis and the appropriate critical region may also be similarly specified.

There exists a formulation of this statistic which corresponds to the Wilcoxon rank sum test for the location problem. Let us rank the positive observations of the two samples together and also rank the negative observations from the two samples together. Let W^+ be the number of pairs (x_i, y_j) in which both are positive and the y_j observation is greater than the x_i observation. Let W^- be the number of pairs (x_i, y_j) in which both are negative and the $-y_j$ observation is greater than the $-x_i$. Also, let n_1^+ and n_2^+ be the number of positive $x's$ and positive $y's$. Then $n_1^- = n_1 - n_1^+$ and $n_2^- = n_2 - n_2^+$ will be the respective number of negative $x's$ and negative

$y's$. Then it is seen that

$$T = \frac{1}{n_1 n_2}[2W^+ + 2W^- - n_1^- n_2^-].$$

The two sample U-statistics results assure that T, after standardization, has $N(0, 1)$ distribution. Under H_0, we have

$$E(T) = 0,$$

and

$$Var(T) = \frac{n}{n_1 n_2}[\frac{1}{3} - \alpha(1 - \alpha)].$$

Thus, for large sample sizes the test could be carried out by comparing the standardized statistic

$$T^* = \frac{T}{\sqrt{\frac{n}{n_1 n_2}(\frac{1}{3} - \alpha(1 - \alpha))}}$$

with the standard normal critical points of the desired level.

7.8 Savage (or Exponential Scores) Test for Positive Valued Random Variables

It is easily seen that the exponential distribution

$$f(x) = \frac{1}{\theta}e^{-\frac{x}{\theta}}, \quad \theta > 0, \quad 0 < x < \infty,$$

has only one parameter θ, which is its expectation as well as its standard deviation (θ^2 being its variance). Hence a change in θ will cause a change in its location as well as dispersion. See Figure 7.6.

In the above example, and in many others, the two probability distributions are ordered according to stochastic dominance. Hence the Mann-Whitney test is suitable here. But [Savage (1956)] has recommended a test based on exponential order statistics for power considerations.

Let X and Y be positive valued random variables with c.d.f's $F(x)$ and $G(x)$, respectively. Let

$$H_0 : F(x) = G(x), \quad 0 \leq x < \infty,$$

and

$$H_0 : F(x) = G(\theta x), \quad 0 \leq x < \infty, \ \theta > 0, \ \theta \neq 1. \tag{7.8.1}$$

The expectation of the ith order statistic from the standard exponential distribution (that is, $F(x) = 1 - e^{-x}$, $x > 0$), is

$$E(X_{(i)}) = \sum_{j=n+1-i}^{n} \frac{1}{j} \ (= a(i) \ (say)).$$

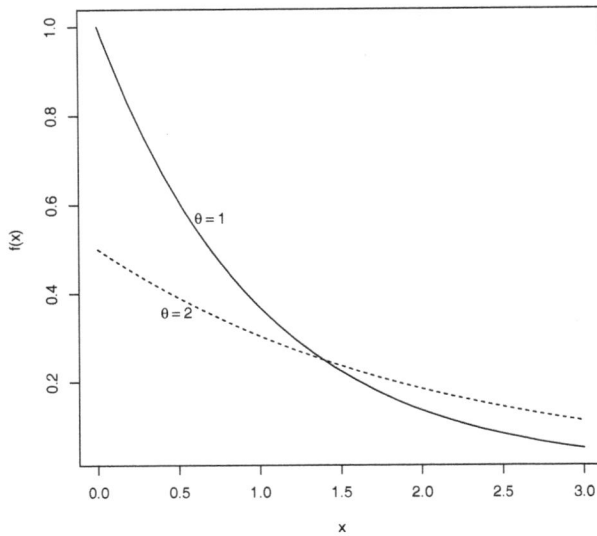

Fig. 7.6   Densities with change in location and scale/dispersion

Now a standard linear rank statistic with these scores is

$$S = \sum_{i=1}^{n_2} a(R_{n_1+i}). \qquad (7.8.2)$$

Here $R_{n_1+1}, R_{n_1+2}, \ldots, R_n$ are the ranks corresponding to the observations of the Y sample in the combined ordering of the observations from the two samples. The linear rank statistic theory tells us that under H_0

$$E(S) = n_2$$

and

$$Var(S) = \frac{n_1 n_2}{n-1}\left(1 - \frac{1}{n}\sum_{i=1}^{n}\frac{1}{i}\right). \qquad (7.8.3)$$

Thus the asymptotic null distribution of the standardized version

$$S^* = \frac{S - n_2}{\sqrt{\frac{n_1 n_2}{n-1}\left(1 - \frac{1}{n}\sum_{i=1}^{n}\frac{1}{i}\right)}}$$

is standard normal. Critcal points with the desired level of significance from the $N(0,1)$ distribution may be used for large sample sizes.

7.9   Kolmogorov-Smirnov Test for the General Two Sample Problem

We have discussed the Kolmogorov-Smirnov test for goodness of fit of a random sample from a given cumulative distribution function. It is based on the supremum of the absolute difference between the prescribed distributions function and the empirical distribution of the data. The same principle is extended to the two sample case. Two empirical distributions are formed for the two random samples and the supremum of the absolute or (signed) difference between them is used as the test statistic.

As is the case in this chapter we have two independent random samples $X_1, X_2, \ldots, X_{n_1}$ and $Y_1, Y_2, \ldots, Y_{n_2}$ from two probability distributions specified by c.d.f's $F(x)$ and $G(x)$, respectively. Consider the testing problem

$$H_0 : F(x) = G(x) \quad \forall \, x$$

against

$$H_1 : \quad F(x) \neq G(x),$$

over a set of non zero probability. This is the two-sided alternative and the two one-sided alternatives are

$$H_{11} : F(x) \geq G(x),$$
$$H_{12} : F(x) \leq G(x),$$

with strict inequality over a set of non zero probability for H_{11} and H_{12}.

By Glivenko-Cantelli theorem one expects the two empirical distribution functions to be close to their (unspecified) true distribution functions, respectively. If the null hypothesis is true, that is, the two c.d.f.'s are the same, then the two empirical distribution functions too are expected to be close to each other. This is illustrated in Figure 7.7.

Consider the two empirical functions defined by

$$F_{n_1}(x) = \frac{1}{n_1}[\# \text{ of } X \text{ observations} \leq x]$$

and

$$G_{n_2}(x) = \frac{1}{n_2}[\# \text{ of } Y \text{ observations} \leq x].$$

The test statistic is based on the difference of these two empirical distribution functions and is defined as

$$D_{n_1,n_2} = \sup_{-\infty < x < \infty} |F_{n_1}(x) - G_{n_2}(x)|. \qquad (7.9.1)$$

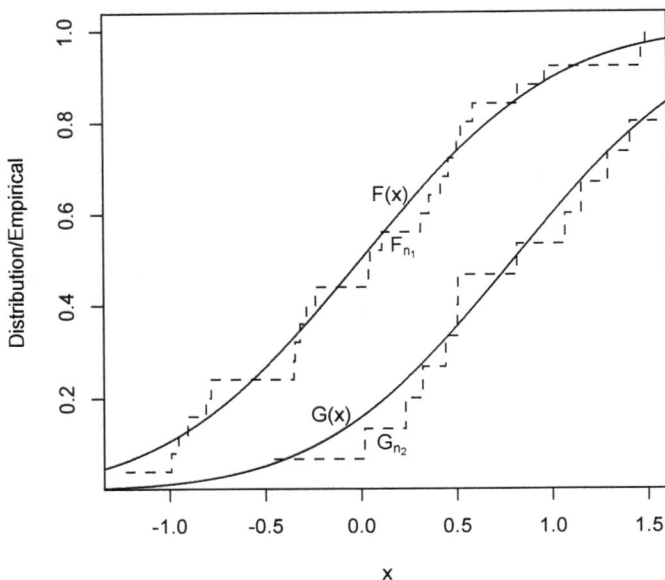

Fig. 7.7 Two distribution functions with empirical distribution functions of respective data sets

Under the null hypothesis the sampling distribution of the statistic D_{n_1,n_2} does not depend upon the common c.d.f. $F(x)$ as long as it is continuous. Combinatorial formula used with counting of rank orders giving different values of the statistic D_{n_1,n_2} will provide its exact null distribution. For increasing n_1, n_2 it soon becomes cumbersome and recourse is taken to the asymptotic distribution provided by [Smirnov (1939)]. Let

$$D^*_{n_1,n_2} = \left(\frac{n_1 n_2}{n_1 + n_2}\right)^{\frac{1}{2}} D_{n_1,n_2}.$$

Then, as $n_1 \to \infty$ and $n_2 \to \infty$,

$$P[D^*_{n_1,n_2} \le s] \to 1 - 2\sum_{r=1}^{\infty}(-1)^{r+1}e^{-r^2 s^2},$$

provided the limit $\frac{n_1}{n_2}$ is neither 0 nor ∞. The 5% and 1% upper critical points for $D^*_{n_1,n_2}$ from the asymptotic distribution are 1.36 and 1.63, respectively. Exact critical points for small sample sizes are available in the literature.

It should be noted that both $F_{n_1}(x)$ and $G_{n_2}(x)$ are jump functions, remaining constant between order statistics. Hence the supremum to be evaluated in (7.9.1) can be obtained as the maximum

$$D_{n_1,n_2} = \max_{1 \leq i \leq n_1+n_2} |F_{n_1}(z_i) - G_{n_2}(z_i)|,$$

where $z_1 \leq z_2 \leq \ldots, \leq z_{n_1+n_2}$ are the combined order statistics of the two random samples.

The one sided versions of the statistic

$$D_{n_1,n_2}^+ = \sup_{-\infty < x < \infty} (F_{n_1}(x) - G_{n_2}(x))$$

and

$$D_{n_1,n_2}^- = \sup_{-\infty < x < \infty} (G_{n_2}(x) - F_{n_1}(x))$$

are the appropriate statistics when the alternatives are suspected to be H_{11} and H_{12}, respectively. The asymptotic null distribution of either of these statistics is especially simple. Asymptotically, both

$$D_{n_1,n_2}^{*+} = \left(\frac{n_1 n_2}{n_1 + n_2}\right)^{\frac{1}{2}} D_{n_1,n_2}^+ \quad \text{and} \quad D_{n_1,n_2}^{*-} = \left(\frac{n_1 n_2}{n_1 + n_2}\right)^{\frac{1}{2}} D_{n_1,n_2}^-$$

have the distribution function $H(x) = 1 - e^{-2x^2}$, $0 < x < \infty$. The exact distribution is discrete and has been tabulated by [Siegel (1956)], [Gail and Green (1976)] and others.

Comment:

The Kolmogorov-Smirnov (two-sided) test is consistent for the entire alternative hypothesis $F(x) \neq G(x)$, for some x. This is because the Glivenko-Cantelli theorem ensures that the two empirical distribution functions tend to the respective distribution function. If there is any non zero difference between them, as the alternative hypothesis implies, it is blown up by the normalizing factor and asymptotically it will be larger than any finite critical point. However, this wide consistency comes at a cost. Studies have shown that the power of the test is often less than that of the Wilcoxon-Mann-Whitney test or the normal scores tests for those alternatives for which they are focused.

7.10   Ranked Set Sampling (RSS) Version of the Wilcoxon-Mann-Whitney Test

As discussed in earlier chapters, the RSS procedure consists of obtaining k units from a population and setting aside the smallest order statistic from these. Then, choose k more units and set aside the second order statistic from these. Continue the process until we obtain k independent order statistics $X_{[1]}, X_{[2]}, \ldots, X_{[k]}$, one each of order 1, order 2, etc. If we repeat this entire procedure n_1 times, then we will have $n_1 k$ observations which are

$$X_{[1]j}, X_{[2]j}, \ldots, X_{[k]j}, \ \ j = 1, 2, \ldots, n_1.$$

These are also independent and have the respective order statistics distributions. We carry out a similar RSS procedures from the second population with groups of q observations and repeat it n_2 times to obtain $n_2 q$ observations

$$Y_{[1]j}, Y_{[2]j}, \ldots, Y_{[q]j}, \ \ j = 1, 2, \ldots, n_2,$$

as the RSS sample.

Let

$$\phi(u) = \begin{cases} 1 & \text{if} \ \ u \geq 0, \\ 0 & \text{if} \ \ u < 0. \end{cases}$$

The RSS version of the Wilcoxon-Mann-Whitney statistic is defined as

$$U_{RSS} = \sum_{s=1}^{q} \sum_{t=1}^{n_2} \sum_{i=1}^{k} \sum_{j=1}^{n_1} \phi(Y_{[s]t} - X_{[i]j}),$$

= number of pairs (X, Y) in the ranked set sample which X is less than or equal to Y,

$$= n_1 n_2 kq \int_{-\infty}^{\infty} F_{n_1 k}^*(t) dG_{n_2 q}^*(t),$$

where $F_{n_1 k}^*$ and $G_{n_2 q}^*$ are the empirical distribution functions of the X and Y RSS data, respectively.

The null hypothesis is $H_0 : F(x) = G(x)$, the equality of the two distribution functions corresponding to the two populations and the alternative hypotheses are

$$H_{11} : F(x) \geq G(x),$$
$$H_{12} : F(x) \leq G(x), \tag{7.10.1}$$

with strict inequalities over sets of non zero probability.

It is shown by [Bohn and Wolfe (1992)] and also by [Öztürk and Wolfe (2000)], that under H_0, $E(U_{RSS}) = \frac{n_1 n_2 k q}{2}$. The entire sampling distribution of U_{RSS}, under H_0, is free of $F(x)$, the common c.d.f. However, since, even under H_0, the data does not consist of i.i.d. observations, but independent order statistics of the specified orders, their rank orders are not equally likely. For example,

$$P[Y(2) < Y(1) < X(2) < X(1)] = \frac{17}{2520},$$

and

$$P[X(1) < X(2) < Y(1) < Y(2)] = \frac{137}{2520}.$$

These probabilities are all distribution free, but involve very tedious calculations.

Asymptotically, using the U-statistic structure, we have

$$[V(U_{RSS})]^{-\frac{1}{2}}[U_{RSS} - E(U_{RSS})]$$

has $N(0,1)$ distribution. If $k = q = 2$, $n_1 = n_2 = n$, then the distribution of

$$\sqrt{2n}[\frac{U_{RSS}}{n^2} - 2] \xrightarrow{d} N(0,1) \text{ as } n \to \infty$$

and critical points of the standard normal distribution may be used. The exact and the asymptotic expressions for the variance are available in [Bohn and Wolfe (1992)].

7.11 Hodges-Lehmann-Sen Estimator for Shift Between Two Continuous Distributions

Let $X_1, X_2, \ldots, X_{n_1}$ be a random sample from the continuous distribution function $F(x)$ and $Y_1, Y_2, \ldots, Y_{n_2}$ be an independent random sample from the distribution $G(x) = F(x - \Delta)$, or $g(x) = f(x - \Delta)$, in terms of the c.d.f.s or the pdfs. First we look at point estimators of Δ. See Figure 7.8.

7.11.1 *Point Estimation of* Δ

Note that

$$\hat{\Delta} = \text{median}_{\substack{1 \leq i \leq n \\ 1 \leq j \leq m}} \{Y_j - X_i\}, \tag{7.11.1}$$

the median of the differences of the observations in the $n_1 n_2$ pairs (X_i, Y_j), provides a point estimator of the shift parameter. In the case of observations

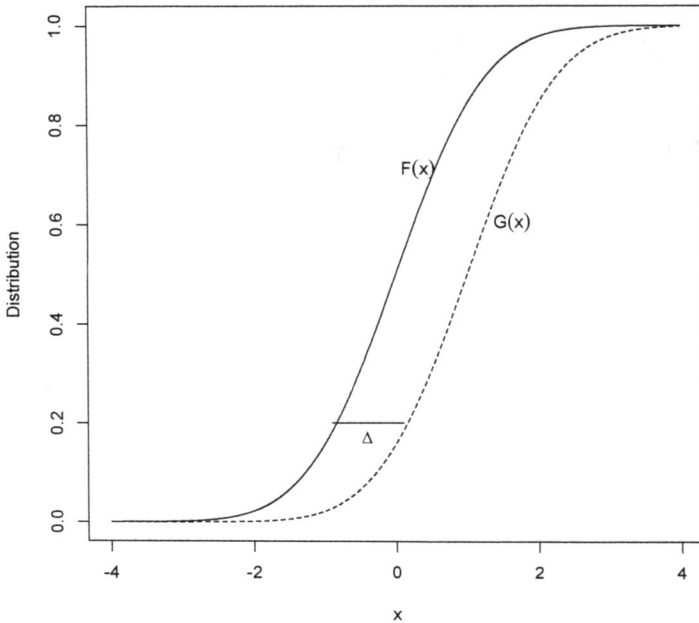

Fig. 7.8   Distributions with location shift

from the normal distribution, $\bar{Y} - \bar{X}$ would be an estimator. In case $n_1 n_2$ is an even number, the average of the two middle most differences would be the estimator of Δ, according to standard convention.

The computation is available in the software R. The motivation is that if $\hat{\Delta}$ is subtracted from the Y observations (or added to X observations) then the value of the Wilcoxon statistic based on these modified values will be closest to its null expectation. This estimator is proposed in [Hodges Jr and Lehmann (1963)] and in [Sen (1963)].

7.11.2   *Confidence Intervals for* Δ

Consider the ordered $n_1 n_2$ differences $(Y_j - X_i)$, $i = 1, 2, \ldots, n_1$, $j = 1, 2, \ldots, n_2$ as $d_{(1)}, d_{(2)}, \ldots, d_{(n_1 n_2)}$. Let $-\infty = d_{(0)}$ and $\infty =$

$d_{n_1 n_2 + 1}$. Then we can partition the real line in $n_1 n_2 + 1$ intervals $(d_{(0)}, d_{(1)}], (d_{(1)}, d_{(2)}], \ldots, (d_{(n_1 n_2)}, d_{(n_1 n_2 + 1)})$. The probability that the ith interval $(d_{(i-1)}, d_{(i)}]$ covers the unknown value of the parameter Δ is the same as the Mann-Whitney statistic U (number of pairs in which Y_j exceeds X_i) takes the value $i - 1$. Hence putting together adequate number of these consecutive intervals will provide the required interval (d_ℓ, d_u) where ℓ and u are chosen such that $\sum_{i=\ell}^{u-1} P(U = i) \simeq 1 - \alpha$. This provides a confidence interval with approximate confidence coefficient $1 - \alpha$ (due to the discontinuities in the distribution of U).

It is convenient to choose ℓ and u in a symmetric manner, that is,

$$\ell + u = n_1 n_2 + 1, \text{ and } \sum_{i=\ell}^{n_1 n_2 + 1 - \ell} P(U = i) \simeq 1 - \alpha.$$

If n_1 and n_2 are reasonably large (say greater than 10 each), then ℓ and u are approximated by

$$\ell = [\frac{n_1 n_2}{2} - z_{\frac{\alpha}{2}} \sqrt{\frac{n_1 n_2 (n_1 + n_2 + 1)}{12}} - 0.5]$$

and

$$u = [\frac{n_1 n_2}{2} + z_{\frac{\alpha}{2}} \sqrt{\frac{n_1 n_2 (n_1 + n_2 + 1)}{12}} - 0.5]$$

where $z_{\frac{\alpha}{2}}$ is the upper $\frac{\alpha}{2}$th quantile of standard normal distribution.

7.12 Exercises

1. Let $X_1, X_2, \ldots, X_n$ be i.i.d. random variables from a continuous distribution $F(x)$ and $Y_1, Y_2, \ldots, Y_m$ from $F(x - \theta)$. Find the exact distribution of

  (i) the Wilcoxon Statistic when $n = 3, m = 4$,
  (ii) the Mann-Whitney Statistic when $n = 4, m = 3$.

  2. Show that under the null hypothesis both the Wilcoxon Statistic and the Mann-Whitney Statistic are symmetric about their respective expected values.

  3. Show that the scores for the Van der Waerden test and Fisher-Yates-Terry-Hoeffding test are asymptotically equivalent.

  4. The following data are life times in years of wheel bearings from two different vendors, with 20 bearings from each. Test whether the variability in the life times is more for vendor II.

Vendor I					Vendor II				
5.76	0.86	2.49	0.55	0.32	0.65	0.12	0.04	0.01	0.16
0.27	0.47	2.70	0.90	2.06	13.23	0.18	0.99	2.12	2.65
8.62	0.54	0.90	2.64	0.23	0.06	10.91	16.12	1.18	0.15
0.74	1.15	1.40	1.57	2.34	14.76	1.48	0.06	16.09	4.00

Chapter 8

THE SEVERAL SAMPLE PROBLEM

8.1 Introduction

In Chapter 7 we have considered the problem of testing the equality of locations of two populations. Now we extend this study to comparing locations of three or more populations. We may regard the procedures in Chapter 7 as alternatives to the two sample t-test when normality of the populations can not be assumed. In this chapter we seek nonparametric alternatives to the F-test as applied to one way classified data from several Normal populations.

Let $F_1(x), F_2(x), \ldots, F_k(x)$ be the distribution functions corresponding to k populations and in general

$$F_i(x) = F(x - \theta_i) \quad i = 1, 2, \ldots, k,$$

where $F_i(x)$ is a continuous (unknown or at least not Normal) distribution function and the differences among the k populations are through the values of the location parameter $\theta_1, \theta_2, \ldots, \theta_k$. If $F(x)$ is normal, then one would use the one-way analysis of variance technique here. In this section we present rank based tests. Let $X_{i1}, X_{i2}, \ldots, X_{in_i}, \quad i = 1, 2, \ldots, k$ be independent and identically distributed random variables with common c.d.f. $F_i(x)$. And if all $\theta_1 = \theta_2 = \cdots = \theta_k$ then all $n = \sum_{i=1}^{k} n_i$ observations would be i.i.d. These k sets of observations will arise when we apply different treatments to the k sets. If all the treatments have the same effect then the null hypothesis

$$H_0 : \theta_1 = \theta_2 = \cdots = \theta_k$$

is true and a reasonable procedure should accept it with a predetermined value α of the probability of Type I error. The alternative says that

$$H_1 : \theta_i \neq \theta_j \text{ for at least one pair } i \neq j.$$

Since all the n observations have a common continuous distribution under H_0, the theory of rank order statistics, as developed in Chapter 6, applies here also. In subsequent sections we propose tests for this problem which are based on rank order statistics.

8.2   The Kruskal-Wallis Test

If we arrange all the $n = \sum_{i=1}^{k} n_i$ observations from the smallest to the largest and assign rank R_{ij} to X_{ij}, the jth, $j = 1, 2, \ldots, n_i$ observation from the ith sample, then the Kruskal-Wallis statistic is defined as

$$H = \frac{12}{n(n+1)} \sum_{i=1}^{k} n_i (\bar{R}_{i.} - \frac{n+1}{2})^2. \tag{8.2.1}$$

Here $\bar{R}_{i.} = \frac{1}{n_i} \sum_{j=1}^{n_i} R_{ij}$, the average rank of the observations from the ith, $i = 1, 2, \ldots, k$ sample. This can be seen to be the rank version of the usual ANOVA defined as (Sum of squares due to the treatment)/(Total sum of squares) except that the total sum of squares here is a non stochastic quantity, being the sum of squares of difference of the numbers $1, 2, \ldots, n$ from their average $(n+1)/2$. A slightly simpler equivalent expression for H is

$$H = \frac{12}{n(n+1)} \sum_{i=1}^{k} n_i \bar{R}_{i.}^2 - 3(n+1). \tag{8.2.2}$$

Under H_0 we expect that the average ranks of k samples will be close to $(n+1)/2$, the overall average of ranks. On the other hand, if the locations of the populations from which these random samples are obtained are different then that should show up in the complete ranking. Then the average ranks will be (atleast in some cases) away from the over all mean rank. This would yield higher values of the statistic H under the alternative H_1 than under H_0. Therefore, the suggested test is

$$\text{Reject } H_0 \text{ if } H > H_\alpha,$$

where H_α is the upper α point of the null distribution of H. The null distribution, being based on the null rank statistics distribution is known and is tabulated for small sample sizes. Many leading software packages include it. The asymptotic null distribution of H as $\min(n_1, n_2, \ldots, n_k)$ tends to infinity, is chi-squared with $k - 1$ degrees of freedom. This is so because of the linear constraint $\sum_{i=1}^{k} n_i \bar{R}_{i.} = \{n(n+1)\}/2$ on the basic random quantities $\bar{R}_{1.}, \bar{R}_{2.}, \ldots, \bar{R}_{k.}$ which are involved in the statistic.

Hence the critical point $\chi^2_{k-1,\alpha}$ should be used for moderately large sample sizes. As in the two sample case, we adjust the statistic H in case there are tied observations in the data. We instead consider

$$H^* = \frac{H}{1 - \frac{\sum_{j=1}^m (t_j^3 - t_j)}{n^3 - n}},\tag{8.2.3}$$

to take care of the reduced variation in the possible values of the statistic H. Here t_i is the number of observations tied in the ith group, there being m such groups. Of course, the distribution of H^* is to be interpreted as its conditional distribution given the number of tied groups and the number of observations in them, that is, $t_1, t_2, \ldots, t_m$.

Example 8.1:

Sample from 3 types of wheat were tested for protein percentage. The data is given in Table 8.1.

Table 8.1

Type I	4.7	4.9	5.5	5.0	5.4	6.1
Type II	6.2	5.6	6.5	5.7		
Type III	5.9	5.6	4.8	7.1	7.0	

Here $n_1 = 6, n_2 = 4, n_3 = 5$, $n = 15$. The rank totals of the 3 samples are $R_1 = 30$, $R_2 = 41.5$, $R_3 = 48.5$, giving the average rank 7.5 to the two observations equal to 5.6.

$$H = \frac{12}{15 \times 16}\left[\frac{(30)^2}{6} + \frac{(41.5)^2}{4} + \frac{(48.5)^2}{5}\right] - 3 \times 16$$
$$= 4.56.$$
$$H^* = 4.569.$$

The 5% critical value of the χ^2_2 distribution is 5.99. The observed value of H is 4.56, which is smaller than the critical value. Hence we cannot reject the null hypothesis that the protein contents in the three varieties of wheat are the same.

In 'R', the function 'kruskal.test' from the package 'stats' implements this test.

8.3   The Bhapkar-Deshpande Test Based on k-plets of Observations

The Mann-Whitney idea of comparing observations in pairs has been extended to k-plets for the k-sample problem by [Bhapkar and Deshpande (1968)].

The null hypothesis is again

$$H_0 : F_1(x) = F_2(x) = \ldots = F_k(x) \ \forall \ x,$$

and independent observations $x_{i1}, x_{i2}, \ldots, x_{in_i}$, $i = 1, 2, \ldots, k$ are available. We form k-plet of observations by selecting one observation from each of the k samples. Obviously, the total number of k-plets is $\prod_{i=1}^{k} n_i$. Define V_{ij} to be the number of k-plet in which the observation from the ith sample is greater than exactly $j - 1$ observations and smaller than the remaining $k - j$ observations in the k-plet. Under the assumption of continuity of random variables, no ties are expected. Consider

$$u_{ij} = \frac{v_{ij}}{(\prod_{i=1}^{k} n_i)}.$$

This is the proportion of such k-plet, that lies in $[0, 1]$, depending on the relative magnitudes of the observations. It is proposed that test statistics be constructed out of these $u'_{ij}s$. We exhibit three such statistics here

(i)

$$V = n(2k - 1)\{\sum_{i=1}^{k} p_i(u_{i1} - \frac{1}{k})^2 - [\sum_{i=1}^{k} p_i(u_{i1} - \frac{1}{k})]^2\}$$

where $n = \sum_{i=1}^{k} n_i$ and $p_i = \frac{n_i}{n}$.

(ii)

$$L = \frac{n(2k - 1)(k - 1)^2 \binom{2k-2}{k-1}}{2k^2[\binom{2k-2}{k-1} - 1]} [\sum_{i=1}^{k} p_i \ell_i^2 - [\sum_{i=1}^{k} p_i \ell_i]^2],$$

where $\ell_i = -u_{i1} + u_{ik}$, $i = 1, 2, \ldots, k$.

(iii)

$$D = \frac{n(2k - 1)(k - 1)^2 \binom{2k-2}{k-1}}{[2(k^2 + (k^2 - 4k + 2)\binom{2k-2}{k-1}]} [\sum_{i=1}^{k} p_i d_i^2 - [\sum_{i=1}^{k} p_i d_i]^2],$$

where $d_i = u_{i1} + u_{ik}$, $i = 1, 2, \ldots, k$.

The functions d_i and D are meaningful if k is at least 3. For $k = 2$ the function d_i is always 1.

Here, the following two alternatives to H_0 are being considered.

A: Assuming identical functional forms for the distribution functions and the location parameters are suspected to be different, and

B: Assuming the same functional forms for the distribution functions and the scale parameters are suspected to be different.

The V and the L test would be efficient for the location alternative A and the V and the D test would be able to detect the scale alternative B.

The V test was proposed by Bhapkar as the k-sample extension of the [Mann and Whitney (1947)] W-test. It was seen to have good asymptotic relative efficiency with respect to the Kruskal-Wallis test for skew distributions such as the exponential distribution. But then it was realized that it has information only in the lower end of the k-plet. Later the L and D statistics were proposed by [Deshpande (1965)] which use information in both sides of the k-plet.

The tests consist in rejecting H_0 at a significance level α if the statistic exceeds the predetermined critical value. [Bhapkar (1961)] and [Deshpande (1965)] have shown that these statistics, under H_0, as each n_i tends to ∞, without n_i/n tending to 0 or 1, have asymptotically χ^2 distribution with $k-1$ degrees of freedom. Hence, for moderately large sample sizes, the upper critical points of the χ^2_{k-1} distribution are used as the cut off points.

Example 8.2: The following table gives the breaking strength in lbs/in^2 of concrete cylinders. The three batches A, B and C differ only in the proportion of coal that the sand used in the concrete contains.

Table 8.2

Batch	Proportion of Coal	Breaking Srength			
A	.00	1410	1670	1475	1505
B	.05	1590	1725	1745	1460
C	.10	1435	1530	1615	1525

Here $k = 3$, $n_i = 4, i = 1, 2, 3$, $n = 12$, $p_i = 1/3$, $i = 1, 2, 3$. We obtain

$$v_{11} = 18, \ v_{13} = 34,$$

$$v_{21} = 39, \ v_{23} = 11,$$

$$v_{21} = 7, \ v_{23} = 19,$$

giving us $\ell_1 = 16/64$, $\ell_2 = -28/64$, $\ell_3 = 12/64$. Then

$$L = 16\left[\sum_{i=1}^{3} p_i \ell_i^2 - \left(\sum_{i=1}^{3} p_i \ell_i\right)^2\right] = 1.54.$$

This is smaller than the upper 5% value of the chi-squared distribution with 2 degrees of freedom, viz, 5.94. Hence there is not enough evidence to reject H_0.

Comments:

(i) [Bhapkar and Deshpande (1968)] showed that the statistic

$$W = \frac{12n}{k^2}[\sum_{i=1}^{k} p_i w_i^2 - [\sum_{i=1}^{k} p_i w_i]^2],$$

where $w_i = \sum_{j=1}^{k}(j-1)u_{ij}$, $i = 1, 2, \ldots, k$, is exactly the same as the Kruskal-Wallis H statistic in case of equal sample sizes, that is, $n_1 = n_2 = \ldots = n_k$. Even if the sample sizes are unequal, W and H are equally efficient in the sense of Pitman ARE . W (or) H is recommended for use when distributions are light tailed and L is recommended in case of heavy tailed distributions. The V test is generally more efficient for exponential type skew distributions and D test is recommended for scale differences among symmetric distributions.

(ii) Later [Deshpande (1972)] defined a statistic based on linear functions $\sum_{j=1}^{k} a_j u_{ij}$ and also derived the optimal coefficients a_j which maximize the asymptotic relative efficiency for a given probability distribution. These tests are not included in this presentation as unlike other nonparametric tests they need a more detailed knowledge of the alternative hypothesis that would be usually unavailable.

(iii) The v_{ij} as defined here are U-statistics with linear constraints $\sum_{i=1}^{k} v_{ij} = \sum_{j=1}^{k} v_{ij} = 1$. The ℓ_i and d_i are their linear functions. The general theory of U-statistics assures the asymptotic normal distribution of the $v'_{ij}s$ and hence asymptotic chi-squared distribution of the V, L, D and W statistics, each with $k - 1$ degrees of freedom.

8.4   Tests to Detect Ordered Alternatives

In the previous two sections we have considered tests of the null hypothesis of equality of k distributions. These are supposed to detect differences on the locations or scale or both in distributions without any known order among the alternatives. However, there are many situations where the analyst has some idea of the possible order among the distributions under the alternative hypothesis. It is possible that the verbal ability of young children is influenced upwards if the parents belong to unskilled worker category, skilled workers or more literate categories such as teachers, etc.

Formally the testing problem may be stated as

$$H_0 : F_1(x) = F_2(x) = \cdots = F_k(x) \ \forall \ x,$$

vs

$$H_1 : F_1(x) \geq F_2(x) \geq \cdots \geq F_k(x),$$

with strict inequality for some x. The alternative hypothesis here describes the situation where the random variables $X_1, X_2, \ldots, X_k$ are progressively stochastically smaller, that is, the observation on them would tend to be progressively smaller. By reversing the inequalities we get the other one sided alternative wherein the random variables become progressively larger.

[Jonckheere (1954)] has proposed the following test procedure. Let $X_{ij}, \ j = 1, 2, \ldots, n_i, \ i = 1, 2, \ldots, k$, be independent random samples from the k populations. Calculate U_{ij} the Mann-Whitney U-statistic for the pair (i, j) random samples and let

$$J = \sum_{i=1}^{k-1} \sum_{j=i+1}^{k} U_{ij}, \tag{8.4.4}$$

where

$$U_{ij} = \sum_{\ell=1}^{n_\ell} \sum_{w=1}^{n_j} \phi(X_{i\ell}, X_{jw})$$

and

$$\phi(X_{i\ell}, X_{jw}) = \begin{cases} 1 \text{ if } X_{i\ell} < X_{jw}, \\ \frac{1}{2} \text{ if } X_{i\ell} = X_{jw}, \\ 0 \text{ otherwise.} \end{cases}$$

Being a linear combination of U-statistics, J has asymptotically normal distribution under both H_0 and H_1. The exact null mean and variance are

$$E_{H_0}(J) = \frac{n^2 - \sum_{i=1}^{k} n_i^2}{4},$$

and

$$Var_{H_0}(J) = \frac{n^2(2n+3) - \sum_{i=1}^{k} n_i^2(2n_i+3)}{72}.$$

Hence, under H_0, the standardized statistic

$$J^* = \frac{J - E_{H_0}(J)}{\sqrt{Var_{H_0}(J)}} \ \overset{d}{\to} \ N(0,1) \text{ as } n \ \to \ \infty.$$

Due to the one-sided nature of the alternative hypothesis J^* will tend to be large under H_1 and H_0 should be rejected if J^* is larger than the upper α critical point. Asymptotocally, the upper α point of the standard normal distribution could be used, for example, for $\alpha = .05$, the critical point is 1.645.

Example 8.3:

The following data gives the yields of sweet potato in 7 seasons across three zones which are supposed to be more and more favourable to this crop.

Table 8.3

	Yields in tons/hectare						
Zone I	4.17	2.30	7.93	4.18	4.18	6.16	5.69
Zone II	5.91	6.03	7.69	5.78	5.78	8.45	6.68
Zone III	6.35	7.52	6.30	8.79	8.78	8.08	11.00

Let μ_I, μ_{II}, μ_{III} be the population means. Test the hypothesis

$$H_0 : \mu_I = \mu_{II} = \mu_{III} \quad \text{against} \quad H_1 : \mu_I \leq \mu_{II} \leq \mu_{III}.$$

We note that $n_1 = n_2 = n_3 = 7$, $n = 21$. Hence

$$J = 126,$$

$$E_{H_0}(J) = \frac{(21)^2 - 3(44)}{4} = \frac{441 - 147}{4} = 73.5,$$

$$Var_{H_0}(J) = \frac{(21)^2(45) - 3(44)(17)}{72} = \frac{19845 - 2499}{72} = 24.1$$

$$J^* = \frac{126 - 73.5}{\sqrt{24.1}} = \frac{52.5}{4.91} = 10.61.$$

Comparing the observed value 10.61 with the standard normal distribution we see that it is so much larger than the critical value even with a very low α, that we reject H_0 and accept H_1 as the reasonable hypothesis.

The function 'jonckheere.test' test from the package 'clinfun' in 'R' performs this test.

8.5   Multiple Comparisons

The purpose of the tests proposed in earlier sections is to see if the k different random samples come from a common population or not. The

rejection of the hypothesis only indicates that there is a significant difference (location or scale) in one or more pairs of the populations. It does
not say which, if any, of the populations or specific order between them are
different from each other. A body of methodology known as 'multiple comparisons' has been developed to address this question. In the parametric
set up with normal distributions techniques proposed by, [Scheffe (1953)],
[Dunnett (1955)], [Tukey (1991)] or those based on Bonferroni inequality
have been in use for a long time and seen to be quite effective. One can
also see [Miller (1981)] and [Benjamini and Hochberg (1995)].

The question of multiple comparisons is usually posed in one of two
ways

(i) many-one comparisons and
(ii) pairwise comparisons.

First we consider the many-one problem.

8.5.1   *Many-one Comparisons*

Let the k populations be distinguished by k different treatments. Let there
be an additional treatment designated as the 'control' treatment. Then
statistical procedures are developed to compare each of the k treatments
with the one control treatment. Hence the nomenclature "many-one". This
is one of the common situations where the currently popular treatment
is designated as the control and the competing k treatments are to be
compared with it. The tests would be one-sided if the purpose is to identify
treatments which are better than the control.

Let μ_0 be the location parameter of the control population from which
n_0 observations $X_{01}, X_{02}, \ldots, X_{0n_0}$ have been obtained. The k populations
which are to be compared with it have yielded $X_{i1}, X_{i2}, \ldots, X_{in_i}$, $i = 1, 2, \ldots, k$. These populations have location parameters $\mu_1, \mu_2, \ldots, \mu_k$, respectively. [Fligner and Wolfe (1982)] have suggested that we should rank
all the $n = n_0 + n_1 + \cdots + n_k$ observations together. Let R_{ij}, $j = 1, 2, \ldots, n_i$, $i = 0, 1, 2, \ldots, k$ be the combined ranks of observations in
$k+1$ samples. Then add the ranks of the observations in the samples other
than the control. The statistic thus is

$$FW = \sum_{i=1}^{k} \sum_{j=1}^{n_i} R_{ij}, \qquad (8.5.5)$$

the ranks of the control sample being excluded in the sum in (8.5.5).

The null hypothesis is

$$H_0 : \mu_0 = \mu_i, \ i = 1, 2, \ldots, k$$

against

$$H_1 : \mu_0 \leq \mu_i, \ i = 1, 2, \ldots, k$$

with at least one strict inequality.

Here the alternative hypothesis does not make any differentiation among the k noncontrol treatment means. So it makes sense to add the ranks of all the observations from the k noncontrol populations.

It is seen, using the rank order results, that

$$E_{H_0}(FW) = \frac{n^*(n+1)}{2},$$

where $n^* = \sum_{i=1}^k n_i$, the total number of observations in the noncontrol samples, and

$$Var_{H_0}(FW) = \frac{n_0 n^*(n+1)}{12}.$$

Then, under H_0, the standardised version

$$FW^* = \frac{FW - \frac{n^*(n+1)}{2}}{\sqrt{\frac{n_0 n^*(n+1)}{12}}},$$

tends to standard normal random variable in distribution as $\min(n_0, n_1, n_2, \ldots, n_k) \to \infty$.

In case of there being g groups of tied observations with t_i observations in the ith group, then the conditional variance is

$$Var_{H_0}(FW|\text{Ties}) = \frac{n_0 n^*}{12}[n + 1 - \frac{\sum_{i=1}^g t_i(t_i - 1)(t_i + 1)}{n(n-1)}].$$

For small sample sizes the test should use the Wilcoxon test critical points for n_0 and n^* observations. For large and even moderate sample sizes the critical points from the standard normal distribution can be used.

Example 8.4: [Hundal (1969)] had carried out a study comparing the productivity of workers with no information (control group) and two groups of workers with some information (two treatment groups). We wish to test whether the control group workers are less productive than the other groups.

Table 8.4

pieces processed

No information group	:	40	35	38	43	44	41
Some information group I	:	38	40	47	44	40	42
Some information group II	:	48	40	45	43	46	44

The ranking giving average rank to tied observations is given in Table 8.5.

Table 8.5

pieces processed

Control group	:	5.5	1	2.5	10.5	13	8
Treatment group I	:	2.5	5.5	17	13	5.5	9
Treatment group II	:	18	5.5	15	10.5	16	13

The sum of ranks of two treatment groups is $\sum_{i=1}^{2} \sum_{j=1}^{6} R_{ij} = 130.5$. Then, $E_{H_0}(FW) = 114$, $Var_{H_0}(FW) = 112.12$.

(The values for the ties corrected expressions when the tied groups are $(38, 38), (40, 40, 40, 40), (43, 43, 43)$ with ranks $(2, 3), (4, 5, 6, 7), (12, 13, 14)$). Thus using the asymptotic expression we get

$$FW^* = \frac{130.5 - 114}{\sqrt{112.12}} = 1.56.$$

This is smaller than, though close to the 5% one sided value of 1.645. Hence we can not reject H_0 at 5%, but should be suspicious of it.

8.5.2 *Multiple Pairwise Comparisons*

Here we come to the problem as it was stated at the beginning of the section. Let there be k populations, say with location parameters $\theta_1, \theta_2, \ldots, \theta_k$. The null and the alternative hypothesis is best specified as

$$H_0 : \theta_i = \theta_j, \ \forall \ i, j$$
$$H_A : \theta_i \neq \theta_j, \ \text{for atleast one pair } (i, j), \ i \neq j.$$

Let $X_{i1}, X_{i2}, \ldots, X_{in_i}, \ i = 1, 2, \ldots, k$ be the k independent random samples from the k populations, respectively. Since we wish to make decisions regarding the equality of the location parameters we rank the observations in each pair of random samples.

Let $R_{i1}, R_{i2}, \ldots, R_{in_i}$ be the ranks of the ith sample when the ith and j^{th} samples are ranked together and

$$W_{ij} = \sum_{\ell=1}^{n_i} R_{i\ell}$$

be their sum. Since this is just the two sample Wilcoxon rank sum based on i^{th} and j^{th} sample observations, one can see that

$$E_{H_0}(W_{ij}) = \frac{n_i(n_i + n_j + 1)}{2},$$

and

$$Var_{H_0}(W_{ij}) = \frac{n_i n_j(n_i + n_j + 1)}{12},$$

Let,

$$W_{ij}^* = \sqrt{2}\frac{W_{ij} - E_{H_0}(W_{ij})}{\sqrt{Var_{H_0}(W_{ij})}}.$$

There are $\frac{k(k-1)}{2}$ such pairwise statistics for $1 \le i < j \le k$.

Multiple comparison procedures are used for declaring if $\theta_i \ne \theta_j$ in some pair (i, j), usually after the rejection of the null hypothesis

$$H_0 : \theta_1 = \theta_2 = \cdots = \theta_k$$

against the general alternative of

$$H_1 : \theta_i \ne \theta_j, \text{ for some } i \ne j$$

by one of the standard k-sample test such as Kruskal-Wallis or the Bhapkar-Deshpande procedure. Once this rejection has occurred further investigation is needed to find out where the differences lie. The rank procedure pioneered by [Steel (1960)], [Steel (1961)] [Dwass (1960)] and [Critchlow and Fligner (1991)] is as follows: (see [Spurrier (2006)]) Obtain a critical point w_α^* such that

$$P_{H_0}[\max_{i<j} |W_{ij}^*| \le w_\alpha^*] = 1 - \alpha.$$

Then compare each W_{ij}^* with w_α^* and declare $\theta_i \ne \theta_j$ if $|W_{ij}^*| > w_\alpha^*$.

This procedure has the experiment wise error rate of α. That is to say, the choice of w_α^* ensures that none of the pair will be declared to have different locations when they are not so with probability $1 - \alpha$.

The critical point w_α^* can be approximated by q_α, the upper α^{th} quantile of the distribution of the range of k independent standard normal random variables, in case $\min(n_1, n_2, \ldots, n_k)$ is large. So the large sample procedure would be

$$\text{Declare } \theta_i \neq \theta_j \text{ if } |W_{ij}^*| > q_\alpha,$$

otherwise declare $\theta_i = \theta_j$.

It is possible (though rare) that even when the preliminary k-sample test rejects H_0 of equality of all θ's, the subsequent multiple comparison test does not find any significant difference in any pair at all. This is because to preserve α as the experiment wise error rate we have used a more stringent criterion for declaring that the difference exists in any given pair.

The asymptotic distribution in case of $n_1 = n_2 = \ldots = n_k$ can be outlined as below.

The asymptotic correlation matrix of $\binom{k}{2} W'_{ij}s$ is the same as that of the $\binom{k}{2}$ differences $Z_i - Z_j$, $i < j$ where the $Z's$ are independent normal $(0,1)$ variables. The factor $\sqrt{2}$ in W_{ij}^* adjusts for the fact that $Z_i - Z_j$ has variance 2. Then the maximum of $\binom{k}{2}$ differences among the $Z's$ is the same as the range of the k independent standard normal random variables. The asymptotic distribution in case of all $n_i's$ are not equal is the same, but needs additional approximation results.

8.6 Comments

(i) In the entire chapter we have described only the techniques based on ranks themselves rather than on functions of ranks such as normal scores, van-der-Waerden or exponential scores, since we wanted to keep the discussion and the working as simple as possible. However, as explained in the chapter on two sample procedures, the ranks may be replaced by more relevant scores which will increase the power of the procedures for the concerned alternatives for specific distribution.

(ii) Using the empirical likelihood approach introduced by Owen (see [Owen (1998)]), a test for equality of k medians in censored data is proposed by [Naik-Nimbalkar and Rajarshi (1997)], which does not need the assumption of equality of the underlying survival functions under the null hypothesis.

(iii) A considerable amount of literature is devoted to the error rates: experiment wise, statement wise, false discovery rate, et al. See, for example, [Tukey (1991)], [Miller (1981)], [Hochberg and Tamhane (2009)], [Benjamini and Hochberg (1995)] and further references.

(iv) R and other popular softwares provide programmes to help calculations and analysis through multiple comparisons, hence the computing effect is no more onerous.

8.7   Exercises

1. Let the verbal ability of three groups of children belonging to three different socio-economic families be represented by the following scores. See whether there is increasing trend in ability as the status improves.

Unskilled	:	12.5	13.5	14.4	15.6	11.3	12.2	10.3	11.1
Skilled	:	14.3	15.2	14.1	18.2	15.3	9.4	13.1	
Teachers	:	14.5	15.8	19.9	12.5	18.4			

2. Use the data of the Example 8.4 from [Hundal (1969)] to carry out the multiple comparison analysis in terms of the location parameters $\theta_1, \theta_2, \theta_3$ of the three categories of workers. Specify the pairs of $\theta's$, if any, in which parameters are declared to be distinct.

Chapter 9

TESTS FOR INDEPENDENCE

9.1 Introduction

In this chapter we consider the problem of testing for independence of a pair of random variables (X, Y). Let $(X_1, Y_1), (X_2, Y_2), \ldots, (X_n, Y_n)$ be a random sample from paired data (X, Y) with a bivariate c.d.f. $F(x, y)$ and unknown correlation coefficient ρ. The sample correlation coefficient r is defined as

$$r = \frac{\sum_{i=1}^{n}(X_i - \bar{X})(Y_i - \bar{Y})}{\sqrt{\sum_{i=1}^{n}(X_i - \bar{X})^2}\sqrt{\sum_{i=1}^{n}(Y_i - \bar{Y})^2}}. \tag{9.1.1}$$

It is usually used as a sample measure of dependence between X and Y.

If (X, Y) has bivariate normal distribution with means μ_1, μ_2, variances σ_1^2, σ_2^2 and correlation ρ ($-\infty < \mu_1, \mu_2 < \infty$, $\sigma_1.\sigma_2 > 0$, $-1 < \rho < 1$), then it is well known that X and Y are independent if and only if ρ, the correlation coefficient between X and Y is 0. Then the test statistic used to test

$$H_0 : X, Y \text{ are independent} \tag{9.1.2}$$

is given by

$$T_1 = \frac{r\sqrt{n-2}}{\sqrt{1-r^2}}.$$

Under the null hypothesis of independence of X and Y the statistic T_1 has t distribution with $n - 2$ degrees of freedom. For testing H_0 against a two-sided alternative of lack of independence one rejects for both large and small values of the statistic T_1.

However, if one wishes to test the hypothesis $\rho = \rho_0$, then one uses the test statistic $T_2 = \frac{1}{2}log_e\frac{1+r}{1-r}$ which, under the null hypothesis of independence and for large n, has Normal distribution with mean $\frac{1}{2}log_e\frac{1+\rho_0}{1-\rho_0}$ and variance $\frac{1}{n-3}$. Here again both large and small values are significant for testing against the two-sided alternative.

However, in the nonparametric context, $\rho = 0$, is neither necessary nor sufficient for the independence of X and Y. Also the statistic T_1 does not have t distribution with $n - 2$ degrees of freedom when the data is from a distribution other than the bivariate normal.

[Hoeffding (1940)] showed that if (X, Y) is a bivariate random vector with $E(X^2) < \infty$, $E(Y^2) < \infty$, then

$$Cov(X, Y) = \int_{-\infty}^{\infty} H(x, y) dx dy, \qquad (9.1.3)$$

where

$$\begin{aligned} H(x, y) &= P[X \leq x, Y \leq y] - P[X \leq x]P[Y \leq y] \\ &= P[X > x, Y > y] - P[X > x]P[Y > y]. \end{aligned} \qquad (9.1.4)$$

In the nonparametric set up there are many alternatives to dependence. A particular alternative for positive dependence which is important and is used in applications is 'Positive Quadrant Dependence' (PQD) [Lehmann (1966)]. (X, Y) is said to be PQD if

$$H(x, y) \geq 0 \ \forall \ x, y \in R; \qquad (9.1.5)$$

with strict inequality over set of nonzero probability. In case of independence the inequality reduces to equality for all x and y.

From (9.1.3) it follows that if (X, Y) is PQD then X and Y are positively correlated. Hence PQD is a stronger measure of dependence than positive correlation. Further if (X, Y) is PQD and $Cov(X, Y)$ is 0, then X, Y are independent.

In the subsequent sections we will study nonparametric procedures for testing the null hypothesis for independence of a pair of random variables given in (9.1.2) with two alternatives in mind -

$$H_1 : X, Y \ \text{are positively correlated,}$$

and

$$H_2 : X, Y \ \text{are PQD.}$$

9.2   Spearman's ρ Statistic

Let us consider the nonparametric set up.   The data consists of $(X_1, Y_1), (X_2, Y_2), \ldots, (X_n, Y_n)$ - n i.i.d. pairs of observations from a bivariate distribution $F(x, y)$. We wish to test the null hypothesis that X, Y

are independent. As mentioned earlier, in general, correlation equal to zero is not equivalent to independence of the pair of random variables.

Let $R_1, R_2, \ldots, R_n$ be the ranks of $X_1, X_2, \ldots, X_n$ among $X's$ alone and $Q_1, Q_2, \ldots, Q_n$ be the ranks of $Y_1, Y_2, \ldots, Y_n$ among $Y's$ alone. Note that, under the hypothesis of independence,

$$P[\underline{R} = \underline{r}, \underline{Q} = \underline{q}] = P[\underline{R} = \underline{r}]P[\underline{Q} = \underline{q}] = \frac{1}{(n!)^2}.$$

In the expression for correlation (9.1.1) replace the pair of observations (X_i, Y_i) by the corresponding pair of ranks (R_i, Q_i). Then $\bar{X}$ and $\bar{Y}$ is replaced by $\frac{n+1}{2}$, the average of ranks $R_i(Q_i)$. We end up with the Spearman's rank correlation coefficient which is defined as

$$\rho_S = \frac{12}{n^3 - n} \sum_{i=1}^{n} (R_i - \frac{n+1}{2})(Q_i - \frac{n+1}{2})$$

$$= \frac{12}{n^3 - n} \sum_{i=1}^{n} R_i Q_i - 3\frac{n+1}{n-1}$$

$$= 1 - \frac{6}{n^3 - n} \sum_{i=1}^{n} (R_i - Q_i)^2.$$

For testing H_0 against H_1, we reject H_0 in favor of H_1 for large values of ρ_S. Under H_0,

$$E(\rho_S) = 0$$

$$Var(\rho_S) = \frac{1}{n-1}.$$

Therefore, as n increases, one can reject H_0 in favor of H_1 for large values of the standardized statistic

$$\rho_S^* = \sqrt{(n-1)}\rho_S.$$

Critical points from standard Normal distribution are used.

Comments:

(i) If both $X's$ and $Y's$ increase together or decrease together in a perfect manner, that is, the largest X_i and Y_i come from the same pair, the second largest X_i and Y_i again come from the same pair and so on, then $R_i = Q_i$, $\forall \; i = 1, 2, \ldots, n$ and $\rho_S = 1$.

(ii) On the other hand, if the $X's$ increase when the $Y's$ decrease and vice versa in a perfect manner, that is the largest X_i and the smallest Y_i come from the same pair, the second largest X_i and the second smallest Y_i again come from the same pair and so on, then $R_i = n - Q_i + 1$, $\forall \; i = 1, 2, \ldots, n$ and $\rho_S = -1$.

(iii) In general, analogous to the bounds for the sample correlation coefficient r, we have $-1 \le \rho_S \le 1$.

(iv) If one needs to test H_0 against the alternative that X and Y are negatively correlated then one rejects for small values of ρ_S or ρ_S^*.

(v) Suppose there are m_1 ties among $X's$ and m_2 ties among $Y's$. Then, instead of ρ_S use the following adjusted version

$$\rho_S^* = \frac{n^3 - n - \sum_{i=1}^n (R_i - Q_i)^2 - 6(t^* + s^*)}{(\sqrt{n^3 - n - 12t^*})(\sqrt{n^3 - n - 12s^*})}, \qquad (9.2.6)$$

where

$$t_i = \text{ number of tied scores at a given rank of } X \text{ observations,}$$

$$s_j = \text{ number of tied scores at a given rank of } Y \text{ observations,}$$

$$t^* = \frac{\sum_{i=1}^{m_1} t_i(t_i^2 - 1)}{12},$$

$$s^* = \frac{\sum_{j=1}^{m_2} s_j(s_j^2 - 1)}{12}.$$

9.3 Kendall's τ Statistic

Another popular nonparametric test statistic for testing independence between a pair of random variables is the Kendall's τ described below. Suppose

$$Sign(u) = \begin{cases} 1 & \text{if} \quad u > 0, \\ 0 & \text{if} \quad u = 0, \\ -1 & \text{if} \quad u < 0. \end{cases}$$

For the same data as in Section 9.2, we consider two pairs of observations (X_i, Y_i) and (X_j, Y_j). These pairs of random variables are called concordant if $sign(X_i - X_j)sign(Y_i - Y_j) > 0$ and the pairs are discordant if $sign(X_i - X_j)sign(Y_i - Y_j) < 0$. Then we calculate the proportion of concordant pairs and subtract from it the proportion of discordant pairs. The Kendall's τ_K test statistic is defined as follows:

$$\tau_K = \frac{1}{n(n-1)} \sum_{1 \le i \ne j \le n} Sign(X_i - X_j)Sign(Y_i - Y_j)$$

$$= \frac{2}{n(n-1)} \sum_{1 \le i < j \le n} Sign(X_i - X_j)Sign(Y_i - Y_j)$$

$$= \frac{2}{n(n-1)} \sum_{1 \le i < j \le n} Sign(R_i - R_j)Sign(Q_i - Q_j).$$

Large values of τ_K are significant for rejecting H_0 in favor of H_1.

Let $\phi_{ij} = Sign(X_i - X_j)Sign(Y_i - Y_j)$. Then, it is easy to see that under H_0, for $j \neq \ell$,

$$P(\phi_{ij} = 1 = \phi_{i\ell}) = P(\phi_{ij} = -1 = \phi_{i\ell}) = \frac{5}{18},$$

$$P(\phi_{ij} = 1, \phi_{i\ell} = -1) = P(\phi_{ij} = -1, \phi_{i\ell} = 1) = \frac{2}{9}.$$

And number of pairs of type $(i, j), (i, \ell)$ with $j \neq \ell$ is $\frac{n(n-1)(n-2)(n-3)}{4}$. It is easy to see that

$$Var(\phi_{ij}) = 1.$$

Combining these facts we get, under H_0,

$$E(\tau_K) = 0,$$
$$Var(\tau_K) = \frac{2(2n+5)}{9n(n-1)}.$$

Comments:

(i) Let A be the number of concordant pairs that is, the number of pairs with ranks in natural order and B be the number of discordant pairs, that is, the number of pairs with ranks in reverse order. The statistic τ_K can also be expressed as

$$\tau_K = \frac{1}{\binom{n}{2}}(A - B) = \frac{2A}{\binom{n}{2}} - 1.$$

Since $A + B = \binom{n}{2}$, it is easy to see that $-1 \leq \tau_K \leq 1$.

(ii) The statistic τ_K can be seen to be a U-statistic estimator for the functional

$$\tilde{\tau}_K = P[(X_i - X_j)(Y_i - Y_j) > 0] - P[(X_i - X_j)(Y_i - Y_j) < 0]. \quad (9.3.7)$$

(iii) One can use the central limit theorem for U-statistics to prove that, under H_0, $\sqrt{n}\tau_K$ has a limiting normal distribution with mean zero and variance $4/9$ for large values of n.

(iv)

$$\tilde{\rho}_S = 3(P[(X_i - X_j)(Y_i - Y_\ell) > 0] - P[(X_i - X_j)(Y_i - Y_\ell) < 0]), \quad (9.3.8)$$

where (X_i, Y_i), (X_j, Y_j) and (X_ℓ, Y_ℓ) are three i.i.d. pairs of observations from $F(x, y)$. Therefore, $\tilde{\rho}_S$, the population version of Spearman's rank correlation ρ_S can also be viewed as a measure of 3 times the difference between the probability of concordance and the probability of discordance, but depending on three pairs of observations, not 2.

(v) Suppose there are m_1 ties among $X's$ and m_2 ties among $Y's$. Then, consider the modified statistic

$$\tau_K^* = \frac{A - B}{(n^* - t^{**})(n^* - s^{**})},$$
(9.3.9)

where,

$$t^{**} = \sum_{i=1}^{m_1} \frac{t_i(t_i - 1)}{2},$$

$$s^{**} = \sum_{j=1}^{m_2} \frac{s_j(s_j - 1)}{2},$$

$$n^* = \frac{n(n - 1)}{2}.$$

Example 9.1: The following table gives the figures for ice cream sales at various temperature. Is there any correlation between the two?

Table 9.1

Temperature	Ice Cream Sales
14.2°	215
16.4°	325
11.9°	185
15.2°	332
18.5°	406
22.1°	522
19.4°	412
25.1°	614
23.4°	544
18.1°	421
22.6°	445
17.2°	408

For the above data $\sum_{i=1}^{12} R_i Q_i = 643$.

Therefore, $\rho_S = 0.951$, $\rho_S^* = 3.254$.

$\tau_K = 0.424$ and $3\sqrt{n}\tau_K/2 = 2.109$.

Hence, there is positive correlation between increase in temperature and ice cream sales.

9.4 Copulas

As a relatively recent development copulas have been used to describe dependence between two random variables. Copulas are multivariate distribution functions whose marginals are uniform distribution over the interval $(0, 1)$. In what follows we restrict ourselves to bivariate copulas. Let $F(x)$, $G(y)$ be the marginal distributions of X, Y and let $F(x, y)$ be the joint distribution of (X, Y). Assume that $F(x), G(y)$ are continuous functions. Because of probability integral transformation $F(X) = U$ and $G(Y) = V$ have uniform distribution on $(0, 1)$. Then, the copula $C(u, v)$ for the pair (X, Y) is given as

$$C(u, v) = P[U \leq u, V \leq v]$$
$$= P[X \leq F^{-1}(u), Y \leq G^{-1}(v)].$$

Following Sklar's theorem we have that there exists a copula $C(u, v)$ such that for all x, y in $[-\infty, \infty]$

$$F(x, y) = C(F(x), G(y)).$$

If $F(x)$ and $G(y)$ are continuous functions, then the copula $C(u, v)$ is uniquely defined. Conversely, the function $F(x, y)$ defined above is a joint distribution function whose marginals are given by F and G, respectively.

Note that for every u, v in $[0, 1]$, we have

$$C(u, 0) = 0 = C(0, v),$$
$$C(u, 1) = u, \; C(1, v) = v.$$

And for all u_1, u_2, v_1, v_2 in $[0, 1]$ such that $u_1 \leq u_2$, $v_1 \leq v_2$

$$C(u_2, v_2) - C(u_2, v_1) - C(u_1, v_2) + C(u_1, v_1) \geq 0.$$

The theory of copulas is useful for generating random samples from multivariate distributions with a given copula.

Three basic copulas of interest are

$$W(u, v) = max(u + v - 1, 0),$$
$$M(u, v) = min(u, v),$$
$$\Pi(u, v) = uv. \qquad (9.4.10)$$

Copulas $W(u, v)$ and $M(u, v)$ come up as bounds to an arbitrary copula $C(u, v)$. Analogous to Frechet bounds for the bivariate distribution function $F(x, y)$, the bounds for the copula function $C(u, v)$ are given below:

$$W(u, v) \leq C(u, v) \leq M(u, v). \qquad (9.4.11)$$

The product copula corresponds to independence of the random variables X and Y.

Next we look at a measure of departure between two copulas.

Suppose (X_1, Y_1) and (X_2, Y_2) are pairs of independent vectors with joint distribution functions $F_1(x, y)$ and $F_2(x, y)$ and copulas $C_1(u, v)$ and $C_2(u, v)$, respectively. $F(x)$ is the marginal distribution of X_1, X_2 and $G(y)$ the marginal distribution of Y_1, Y_2, respectively. Then difference between the probability of concordance and probability of discordance is

$$\tilde{\tau}(C_1, C_2) = P[(X_i - X_j)(Y_i - Y_j) > 0] - P[(X_i - X_j)(Y_i - Y_j) < 0]$$
$$= 4 \int\int_{[0,1]X[0,1]} C_2(u, v)dC_1(u, v) - 1. \qquad (9.4.12)$$

Note that $\tilde{\tau}(C_1, C_2)$ is symmetric in its arguments.

When $C_1(u, v) = C_2(u, v) = C(u, v)$, then $\tilde{\tau}(C_1, C_2)$ is given as

$$\tilde{\tau}(C, C) = 4 \int\int_{[0,1]X[0,1]} C(u, v)dC(u, v) - 1 = 4E(C(U, V)) - 1. \quad (9.4.13)$$

Hence $\tilde{\tau}(C, C)$ is a linear function of the expectation of $C(U, V)$ where U, V are dependent uniform $(0, 1)$ random variables. Notice that the population version of Kendall's τ_K is

$$\tilde{\tau}(C, C) = \tilde{\tau}_K.$$

Similarly, the population version of Spearman's ρ_S, that is , $\tilde{\rho}_S$ is given as

$$\tilde{\rho}_S = 12 \int\int_{[0,1]X[0,1]} [C(u, v) - uv]dudv$$
$$= 12 \int\int_{[0,1]X[0,1]} uvdC(u, v) - 3$$
$$= 3\tilde{\tau}(C, \Pi),$$

where $\Pi(u, v)$ is the independent copula given in (9.4.10).

9.4.1 *Empirical Copulas*

Let $X_{(1)}, X_{(2)}, \ldots, X_{(n)}$ and $Y_{(1)}, Y_{(2)}, \ldots, Y_{(n)}$ be the order statistics corresponding to $X_1, X_2, \ldots, X_n$ and $Y_1, Y_2, \ldots, Y_n$, respectively. Then, a nonparametric estimator of the copula function is the empirical copula [Deheuvels (1979)] defined as

$$C_n(\frac{i}{n}, \frac{j}{n}) = \frac{\text{Number of pairs } (X_k, Y_k) \text{ with } X_i \leq X_{(i)}, Y_i \leq Y_{(j)}}{n}.$$
$$(9.4.14)$$

Let,

$$c_n(\frac{i}{n}, \frac{j}{n}) = \begin{cases} \frac{1}{n} & \text{if } (X_{(i)}, Y_{(j)}) \text{ is an element of the sample,} \\ 0 & \text{otherwise.} \end{cases}$$

The relationship between $c_n(\frac{i}{n}, \frac{j}{n})$ and $C_n(\frac{i}{n}, \frac{j}{n})$ is given below:

$$c_n(\frac{i}{n}, \frac{j}{n}) = C_n(\frac{i}{n}, \frac{j}{n}) - C_n(\frac{i-1}{n}, \frac{j}{n}) - C_n(\frac{i}{n}, \frac{j-1}{n}) + C_n(\frac{i-1}{n}, \frac{j-1}{n}),$$

and

$$C_n(\frac{i}{n}, \frac{j}{n}) = \sum_{\ell=1}^{i} \sum_{\ell'=1}^{j} c_n(\frac{\ell}{n}, \frac{\ell'}{n}).$$

9.4.2 Tests Based on Copulas

In this section we look at tests for independence against the alternative H_2 based on copulas studied by [Deheuvels (1979)]. In terms of copulas independence is equivalent to

$$C(u, v) = uv, \quad 0 \le u, v \le 1 \tag{9.4.15}$$

and the pair (X, Y) is said to be PQD if

$$C(u, v) \ge uv, \quad 0 \le u, v \le 1. \tag{9.4.16}$$

Hence one measure of departure from independence is given by

$$\gamma = \int_0^1 \int_0^1 [C(u, v) - uv] du dv. \tag{9.4.17}$$

A test for testing H_0 versus positive dependence can be based on γ_n, an empirical estimate of γ. But $12\gamma = \tilde{\rho}_S$. Hence, a test based on γ_n is equivalent to the test based on Spearman's statistic ρ_S.

Tests of independence based on copulas have been studied by [Deheuvels (1979)]. He proposed a Kolmogorov type supremum statistic for testing for departure from independence. Consider

$$T_n(1) = \max_{1 \le i, j \le n} |C_n(\frac{i}{n}, \frac{j}{n}) - \frac{ij}{n^2}|. \tag{9.4.18}$$

Large values of the statistic are significant. The small sample exact null distribution of $n^2 T_n(1)$ was tabulated.

Another test was proposed by [Genest and Rémillard (2004)]. Consider

$$\tilde{C}_n(u, v) = C_n(u, v) - C_n(u, 1) C_n(1, v).$$

Let

$$B_n = \int_0^1 \int_0^1 [\tilde{C}_n(u,v)]^2 dudv = \frac{1}{n} \sum_{i=1}^n \sum_{j=1}^n D_n(R_i, R_j) D_n(Q_i, Q_j), \quad (9.4.19)$$

where

$$D_n(s,t) = \frac{2n+1}{6n} + \frac{s(s-1)}{2n(n+1)} + \frac{t(t-1)}{2n(n+1)} - \frac{max(s,t)}{n+1}.$$

Large values of the statistic are significant. Tables for asymptotic critical values are given in [Genest and Rémillard (2004)]. Asymptotic properties of B_n are studied in [Genest *et al.* (2007)].

9.5   Positive Quadrant Dependence

In this section we consider two measures of quadrant dependence and their relationship with $\tilde{\rho}_S$ and $\tilde{\tau}_K$.

The first measure of quadrant dependence where the average is taken with respect to F and G, the marginal distribution functions of X and Y is

$$\int_{-\infty}^{\infty} \int_{-\infty}^{\infty} H(x,y) dF(x) dG(y) \quad (9.5.20)$$

$$= \int \int_{[0,1]X[0,1]} [C(u,v) - uv] dudv$$

$$= \int \int_{[0,1]X[0,1]} uv dC(u,v) - 3$$

$$= \frac{1}{12} \tilde{\rho}_S. \quad (9.5.21)$$

Another measure of quadrant dependence where the average is taken with respect to $F(x,y)$, the joint distribution functions of X and Y is

$$\int_{-\infty}^{\infty} \int_{-\infty}^{\infty} H(x,y) dF(x,y) \quad (9.5.22)$$

$$\int_{-\infty}^{\infty} \int_{-\infty}^{\infty} [F(x,y) - F(x)G(y)] dF(x,y)$$

$$= \int \int_{[0,1]X[0,1]} [C(u,v) - uv] dC(u,v)$$

$$= \frac{1}{12} [3\tilde{\tau}_K - \tilde{\rho}_S]. \quad (9.5.23)$$

[Nelsen (1992)], [Tchen (1980)] and [Lehmann (1966)] showed that if (X, Y) are PQD, then,

$$\tilde{\rho}_S \geq 0,$$
$$\tilde{\tau}_K \geq 0,$$
$$3\tilde{\tau}_K - \tilde{\rho}_S \geq 0. \tag{9.5.24}$$

Hence the tests based on Spearman's ρ_S and τ_K used for testing H_0 versus H_1 can also be used for testing H_0 against the PQD alternative H_2. In particular we have

$$3\tilde{\tau}_K \geq \tilde{\rho}_S.$$

Comments:

(i) [Kochar and Deshpande (1988)] proposed tests based on U-statistics for testing H_0 vesrus H_2.

(ii) In some cases one observes a pair of random variables (X, Y), where X is a continuous random variable and Y is discrete. Here X could be the failure time of an individual and Y the cause of failure or X is the number of years of marriage and Y and indicator variable denoting 1 in case of death and 0 if there is a divorce. [Dewan *et al.* (2004)] proposed tests based on U-statistic for testing H_0 vs H_1 and H_2 for such data.

(iii) Suppose $F_1(x, y)$ and $F_2(x, y)$ are two bivariate distribution functions with marginal distribution functions F and G. [Lai and Xie (2000)] defined the concept of more PQD for comparing $F_1(x, y)$ and $F_2(x, y)$. $F_2(x, y)$ is more concordant than $F_1(x, y)$ if

$$F_1(x, y) \leq F_2(x, y), \quad \forall x, y \in R.$$

Consider the following copula given by [Lai and Xie (2000)].

$$C_\rho(u, v) = uv + \rho u^b v^b (1 - u)^a (1 - v)^a, \quad a, b \geq 1, \ 0 < \rho < 1.$$

Then for $\rho_1 < \rho_2$ we have C_{ρ_2} is more concordant than C_{ρ_1}.

(iv) It is easy to see that

$$\tilde{\tau}(M, M) = 1,$$
$$\tilde{\tau}(M, \pi) = 1/3,$$
$$\tilde{\tau}(M, W) = 0,$$
$$\tilde{\tau}(W, \pi) = -1/3,$$
$$\tilde{\tau}(W, W) = -1,$$
$$\tilde{\tau}(\pi, \pi) = 0.$$

9.6   Exercises

1. Prove the following:

(a) $-1 \leq \rho_S \leq 1$,
(b) $-1 \leq \tau_K \leq 1$.

2(a). The statistic τ_K can be seen to be a U-statistic estimator for the functional

$$\tilde{\tau_K} = P[(X_i - X_j)(Y_i - Y_j) > 0] - P[(X_i - X_j)(Y_i - Y_j) < 0]. \quad (9.6.25)$$

(b). Find the asymptotic distribution under the hypothesis of independence of $\sqrt{n}\tau_K$ as $n \to \infty$ using the fact that τ_K is a U-statistic.

Chapter 10

NONPARAMETRIC DENSITY ESTIMATION

10.1 Introduction

Let $X_1, X_2, \ldots, X_n$ be independent and identically distributed random variables with density function $f(x)$. If the density function $f(x)$ were of a known parametric form, for example, Normal (θ, σ^2) with

$$f(x) = \frac{1}{\sqrt{2\pi}\sigma} exp\left\{-\frac{1}{2}\frac{(x-\theta)^2}{\sigma^2}\right\}, \quad -\infty < x, \ \theta < \infty, \ \sigma > 0,$$

then one could estimate the unknown mean θ and variance σ^2 using standard procedures for parametric estimation. Then, one can plug these estimators in the expression for the unknown density function to get its estimator.

In this chapter we will look at nonparametric estimators of the density function $f(x)$ for a fixed value of x. At the outset it should be noted that one can not obtain a non-negative unbiased estimator of the density function.

Suppose there exists a non-negative unbiased estimator $S_n(x)$ of $f(x)$ such that

$$E(S_n(x)) = f(x), \ \forall \ x.$$

By Fubini's Theorem, we have

$$E[\int_a^b S_n(x)dx] = \int_a^b E[S_n(x)]dx$$
$$= \int_a^b f(x)dx$$
$$= F(b) - F(a).$$

Also, we have,

$$E[F_n(b) - F_n(a)] = F(b) - F(a),$$

163

where $F_n(x)$ is the empirical distribution function discussed in Chapter 4. Since the order statistics $(X_{(1)}, X_{(2)}, \ldots, X_{(n)})$ constitute a complete sufficient statistics and $F_n(b) - F_n(a)$ is a function of the order statistics, it follows that

$$F_n(b) - F_n(a) = \int_a^b S_n(x)dx \quad a.e.$$

But this is not possible as $F_n(x)$ is not an absolutely continuous distribution function. Hence, $f(x)$ can not have a pointwise non-negative unbiased estimator.

10.2   The Histogram

The earliest and the simplest estimator of the density function is the histogram. Let x_0 be the origin and consider the intervals $[x_0 + kh, x_0 + (k + 1)h]$, $k = \ldots, -2, -1, 0, 1, 2, \ldots$. Then, the histogram is defined as

$$\hat{f}(x) = \frac{1}{nh}(\text{Number of } X_i's \text{ in the same bin as } x). \tag{10.2.1}$$

The intervals constitute the bins of the histogram. Figures 10.1-10.4, 10.5-10.8, 10.9-10.13 give the plots for standard Normal density function and the histogram for various choices of the sample size, the origin and the bandwidth. The standard normal density function is in green in each of the figures.

Comments:

(i) Figures 10.1-10.4 show the histogram and the density function of a standard Normal distribution with origin 0 and the bin width 0.5. The various choices of sample size are $n = 50, 100, 500, 1000$. From the figures it is clear that for small sample sizes the histogram estimator is not close to the unknown density function, but improves with increase in sample size.

(ii) Figures 10.5-10.8 show the histograms for a random sample of size 2000 from standard Normal distribution with bandwidth 0.5 and origin varying from $-1.0, -0.5, 0.5, 1.0$.

(iii) Figures 10.9-10.13 show the histograms for a random sample of size 2000 from standard Normal distribution with origin 0.0 and varying bandwidth $0.1, (0.2), \ldots, 0.9$.

True and Estimated Density

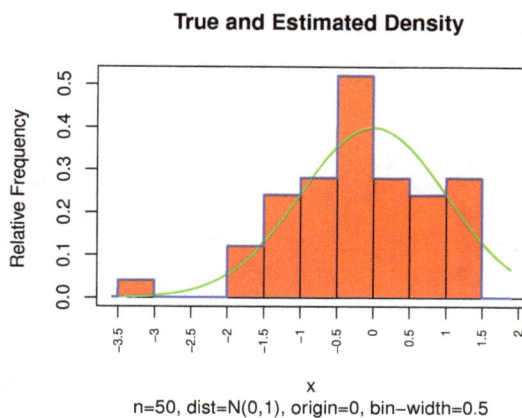

Fig. 10.1   (Histogram - $n = 50$)

True and Estimated Density

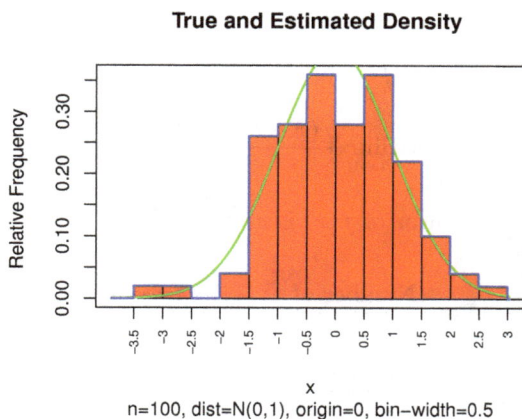

Fig. 10.2   (Histogram - $n = 100$)

Clearly, the histogram estimator $\hat{f}(x)$ depends on the origin x_0 and the choice of the bandwidth which controls the smoothing. This makes the estimator $\hat{f}(x)$ very subjective. Besides $\hat{f}(x)$ is discontinuous and is not appropriate for estimation of derivative of the density function $f(x)$.

True and Estimated Density
n=500, dist=N(0,1), origin=0, bin-width=0.5

True and Estimated Density
n=1000, dist=N(0,1), origin=0, bin-width=0.5

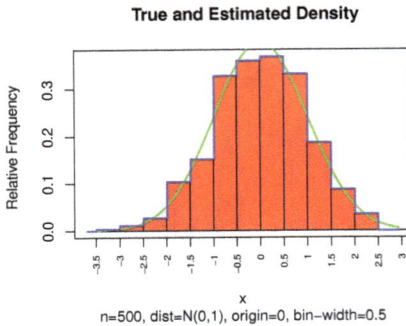

Fig. 10.3   (Histogram - $n = 500$)   Fig. 10.4   (Histogram - $n = 1000$)

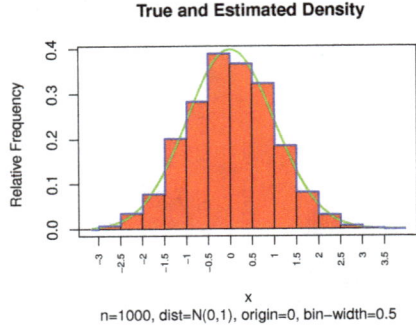

10.3   The Naive Estimator

The density function $f(x)$ is the derivative of the distribution function $F(x)$. Therefore, we can write $f(x)$ as

$$f(x) = \lim_{h \to 0} \frac{1}{2h} P[x - h < X \le x + h]$$

$$= \lim_{h \to 0} \frac{F(x + h) - F(x)}{2h}.$$

Hence the naive estimator of $f(x)$ is given by,

$$\tilde{f}(x) = \frac{(\text{Number of } X_i's \text{ in } (x - h, x + h])}{2nh}$$

$$= \frac{F_n(x + h) - F_n(x - h)}{2nh}.$$

One can rewrite $\tilde{f}(x)$ as

$$\tilde{f}(x) = \frac{1}{n} \sum_{i=1}^{n} \frac{1}{h} w(\frac{x - X_i}{h}), \tag{10.3.2}$$

where

$$w(x) = \begin{cases} \frac{1}{2} & \text{if } x \in [-1, 1), \\ 0 & \text{otherwise.} \end{cases} \tag{10.3.3}$$

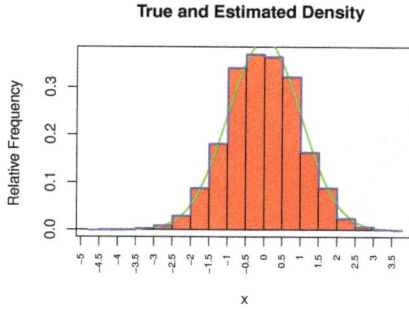

Fig. 10.5   (Histogram - Origin -1.0)

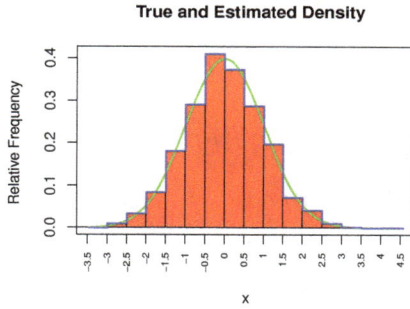

Fig. 10.6   (Histogram - Origin $-.5$)

Naive estimator, in some sense, is the generalization of the histogram estimator where every observed value is the centre of the interval/bin. Thus, it takes care of the subjectivity in choosing an origin in the case of histogram estimators. The smoothness of the estimator is determined by the choice of the bin width. The larger the bin width, the smoother the estimator. But too large bin widths can lead to flat estimators.

It is easy to see that

$$E[\tilde{f}(x)] = \frac{F(x+h) - F(x-h)}{2h}, \qquad (10.3.4)$$

$$Var[\tilde{f}(x)] = \frac{1}{nh^2} Var[K(\frac{x - X_1}{h})]$$

$$= \frac{1}{4nh^2}[F(x+h) - F(x-h) - (F(x+h) - F(x-h))^2],$$

$$(10.3.5)$$

True and Estimated Density

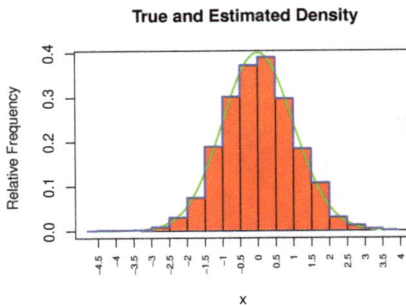

True and Estimated Density

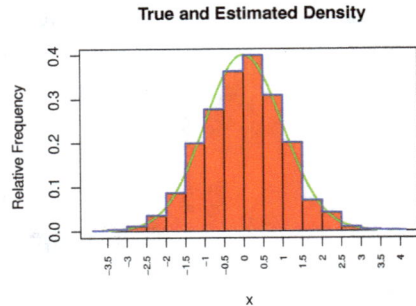

Fig. 10.7   (Histogram - Origin .5)   Fig. 10.8   (Histogram - Origin 1.0)

$$Cov[\tilde{f}(x), \tilde{f}(y)] = \frac{1}{4h^2} Cov(F_n(x+h) - F_n(x-h), F_n(y+h) - F_n(y-h))$$

$$= \frac{1}{4h^2}[(F((x+h) \wedge (y+h)) - F(x+h)F(y+h))$$

$$-(F((x+h) \wedge (y-h)) - F(x+h)F(y-h))$$

$$-(F((x-h) \wedge (y+h)) - F(x-h)F(y+h))$$

$$+(F((x-h) \wedge (y-h)) - F(x-h)F(y-h))]. \quad (10.3.6)$$

The mean squre error (MSE) of the naive estimator $\tilde{f}(x)$ is given by

$$E[\tilde{f}(x) - f(x)]^2 = Var[\tilde{f}(x)] + [E(\tilde{f}(x)) - f(x)]^2$$

$$= \frac{1}{4nh^2}[F(x+h) - F(x-h) - [F(x+h) - F(x-h)]^2]$$

$$+[\frac{1}{2h}[F(x+h) - F(x-h)] - f(x)]^2. \quad (10.3.7)$$

Assuming that the first three derivatives of $f(x)$ exist and the second derivative $f^{(2)}(x) \neq 0$, consider the Taylor Series expansion of $F(x+h)$ and $F(x-h)$, we get,

$$F(x+h) - F(x-h) = 2hf(x) + \frac{1}{3}f^{(2)}(x)h^3 + O(h^4). \quad (10.3.8)$$

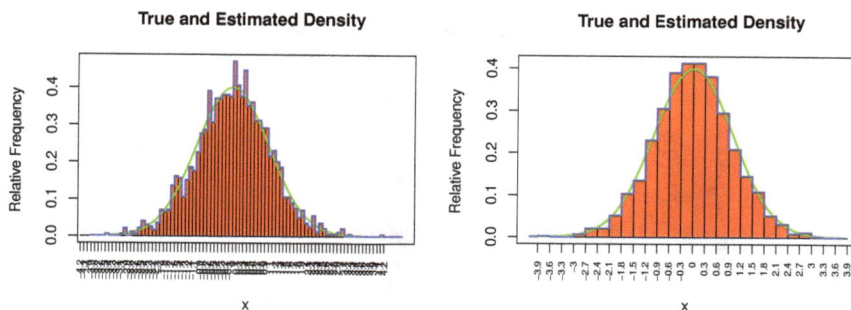

Fig. 10.9   (Histogram - Bandwidth 0.1)   Fig. 10.10   (Histogram - Bandwidth 0.3)

Using (10.3.8), the bias of the naive estimator $\tilde{f}(x)$ is given by,

$$E[\tilde{f}(x) - f(x)] = f(x) + \frac{1}{6}f^{(2)}(x)h^2 + O(h^3)$$

$$\approx \frac{h^2 f^{(2)}(x)}{6}, \text{ as } h \to 0. \tag{10.3.9}$$

$$Var[\tilde{f}(x)] = \frac{1}{4nh^2}[(2hf(x) + \frac{1}{3}f^{(2)}(x)h^3 + O(h^4))$$

$$-(2hf(x) + \frac{1}{3}f^{(2)}(x)h^3 + O(h^4))^2]$$

$$\approx \frac{f(x)}{2nh}, \text{ as } h \to 0. \tag{10.3.10}$$

Substituting (10.3.9) and (10.3.10) in (10.3.7), we get the MSE of $\tilde{f}(x)$ is

$$E[\tilde{f}(x) - f(x)]^2 \approx \frac{f(x)}{2nh} + \frac{h^4}{36}(f^{(2)}(x))^2 + O(h^4 + \frac{1}{nh}). \tag{10.3.11}$$

Observe that the bias tends to zero but the variance tends to infinity as $h \to 0$. Hence choose h such that

$$h^4 \approx \frac{1}{nh}. \tag{10.3.12}$$

True and Estimated Density

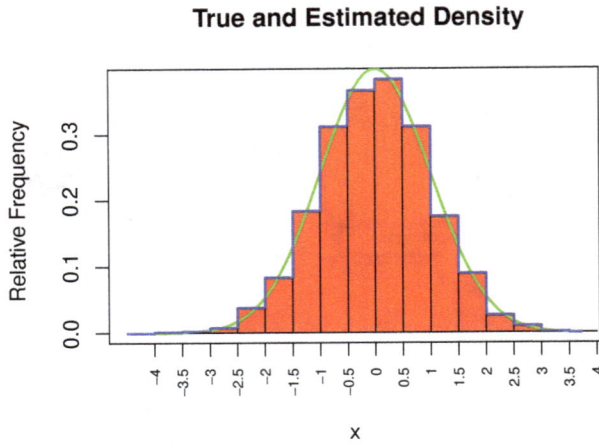

Fig. 10.11   (Histogram - Bandwidth 0.5)

Let

$$h = kn^{-\alpha}, \; \alpha > 0.$$

Then, from (10.3.12), it follows that

$$h = n^{-\frac{1}{5}},$$

and MSE of $\tilde{f}(x)$ is

$$\frac{f(x)}{2k}n^{-\frac{4}{5}} + k^4 n^{-\frac{4}{5}} \frac{(f^{(2)}(x))^2}{36}. \qquad (10.3.13)$$

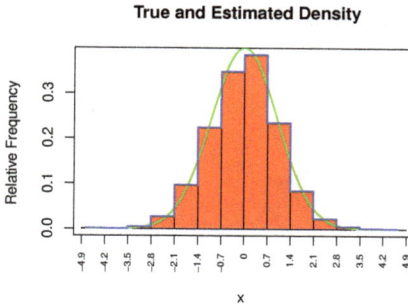

Fig. 10.12 (Histogram - Bandwidth 0.7)   Fig. 10.13 (Histogram - Bandwidth 0.9)

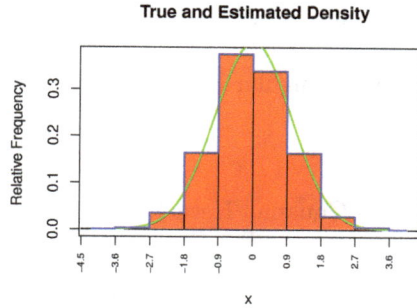

Minimizing (10.3.13) with respect to k, the optimal choice of k is

$$k = [\frac{\frac{9}{2} f(x)}{|f^{(2)}(x)|^2}]^{\frac{1}{5}}. \tag{10.3.14}$$

Substituting in (10.3.13), the optimal MSE of $\tilde{f}(x)$ is given by

$$E[\tilde{f}(x) - f(x)]^2 \approx \frac{5}{4} 9^{-\frac{1}{5}} 2^{-\frac{4}{5}} (f(x))^{\frac{4}{5}} |f^{(2)}(x)|^{\frac{2}{5}} n^{-\frac{4}{5}}. \tag{10.3.15}$$

10.4 Kernel Estimators

The naive estimator leads us to a class of estimators of the form

$$f_n(x) = \frac{1}{nh_n} \sum_{i=1}^{n} K(\frac{x - X_i}{h_n}), \tag{10.4.16}$$

where the bandwidth $h_n \to 0$ as $n \to \infty$. $f_n(x)$ is the kernel estimator of the unknown density function $f(x)$. The kernel $K(.)$ is a symmetric density function defined on the real line satisfying the following conditions:

$$A1 : \sup_{-\infty < x < \infty} K(x) \le M,$$

$$A2 : |x| K(x) \to 0 \text{ as } |x| \to \infty,$$

$$A3 : \int_{-\infty}^{\infty} x^2 K(x) dx < \infty. \tag{10.4.17}$$

Examples of kernels satisfying the conditions given above are:

(i) Triangular Kernel:

$$K(x) = \begin{cases} 1 - |x| & \text{if } |x| < 1, \\ 0 & \text{otherwise.} \end{cases}$$

(ii) Gaussian Kernel:

$$K(x) = \frac{1}{\sqrt{2\pi}} exp - (\frac{1}{2}x^2), \quad -\infty < x < \infty.$$

(iii) Biweight Kernel:

$$K(x) = \begin{cases} \frac{15}{16}(1 - x^2)^2 & \text{if } |x| < 1, \\ 0 & \text{otherwise.} \end{cases}$$

(iv) Rectangular Kernel:

$$K(x) = \begin{cases} \frac{1}{2} & \text{if } |x| < 1, \\ 0 & \text{otherwise.} \end{cases}$$

Note that the naive estimator $\tilde{f}(x)$ is a special case of the kernel type estimator with rectangular kernel.

$$\begin{aligned} E[f_n(x)] &= \int_{-\infty}^{\infty} \frac{1}{h_n} K(\frac{x-y}{h_n}) f(y) dy \\ &\to f(x) \int_{-\infty}^{\infty} K(y) dy \\ &= f(x), \end{aligned} \qquad (10.4.18)$$

for a continuous function $f(x)$.

$$\begin{aligned} Var[f_n(x)] &= \frac{1}{n} Var[\frac{1}{h_n} K(\frac{x-X_1}{h_n})] \\ &\leq \frac{1}{n} E[\frac{1}{h_n} K(\frac{x-X_1}{h_n})]^2 \\ &= \frac{1}{nh_n} \int_{-\infty}^{\infty} \frac{1}{h_n} K^2(\frac{x-y}{h_n}) f(y) dy \\ &\to \frac{1}{nh_n} f(x) \int_{-\infty}^{\infty} K^2(y) dy \quad \text{as } n \to \infty. \qquad (10.4.19) \end{aligned}$$

Comments:

(i) From (10.4.18) it follows that the kernel type estimator $f_n(x)$ is asymptotically unbiased estimator of the unknown density function $f(x)$.

(ii) Further, since (10.4.19) holds, $f_n(x)$ is a pointwise consistent estimator for $f(x)$.

(iii) If $K(.)$ is a function of bounded variation and the series $\sum_{n=1}^{\infty} exp(-\gamma n h_n^2)$ converges for every $\gamma > 0$, then

$$V_n = \sup_{-\infty < x < \infty} |f_n(x) - f(x)| \to 0 \text{ with probability 1 as } n \to \infty,$$

iff density $f(x)$ is uniformly continuous.

This result gives a necessary and sufficient condition for uniform consistency with probability 1.

Next we look at the central limit theorem for $f_n(x)$ for a fixed x. Note that $f_n(x)$ can be expressed as

$$f_n(x) = \frac{1}{n} \sum_{i=1}^{n} Z_{ni}(x), \qquad (10.4.20)$$

where

$$Z_{ni}(x) = \frac{1}{h_n} K\left(\frac{x - X_i}{h_n}\right).$$

Thus, $f_n(x)$, for a fixed x, is the average of an array of random variables $\{Z_{ni}, 1 \le i \le n\}$. Then, if $f(x)$ is a continuous density function and $K(x)$ is a symmetric kernel satisfying the assumptions (A1) - (A3), we have,

$$\frac{f_n(x) - E(f_n(x))}{\sqrt{Var(f_n(x))}} \xrightarrow{D} N(0,1), \text{ as } n \to \infty, \qquad (10.4.21)$$

(where $h_n \to 0$ and $nh_n \to \infty$ as $n \to \infty$.)

A sufficient condition for (10.4.21) to hold is that for some $\delta > 0$,

$$\frac{E|Z_{n1}(x) - E(Z_{n1}(x))|^{2+\delta}}{n^{\frac{\delta}{2}}(Var(Z_{n1}(x)))^{1+\frac{\delta}{2}}} \to 0 \text{ as } n \to \infty. \qquad (10.4.22)$$

One can see that

$$E|Z_{n1}(x)|^{2+\delta} = \int_{-\infty}^{\infty} |\frac{1}{h_n} K\left(\frac{x - y}{h_n}\right)|^{2+\delta} f(y) dy$$

$$\simeq \frac{1}{h_n^{1+\delta}} f(x) \int_{-\infty}^{\infty} |K(y)|^{2+\delta} dy. \qquad (10.4.23)$$

$$Var(Z_{n1}(x)) = Var(\frac{1}{h_n}K(\frac{x - X_i}{h_n}))$$

$$= \frac{1}{nh_n^2}\int_{-\infty}^{\infty} K^2(\frac{x-y}{h_n})f(y)dy - \frac{1}{nh_n^2}[\int_{-\infty}^{\infty} K(\frac{x-y}{h_n})f(y)dy]^2$$

$$= \frac{1}{nh_n^2}\int_{-\infty}^{\infty} K^2(\frac{x-y}{h_n})f(y)dy - \frac{1}{n}[f(x) + Bias]^2$$

$$\simeq \frac{1}{nh_n}f(x)\int_{-\infty}^{\infty} K^2(t)dt + O(\frac{1}{n})$$

$$\simeq \frac{1}{nh_n}f(x)\int_{-\infty}^{\infty} K^2(t)dt. \qquad (10.4.24)$$

Using C_r inequality, (10.4.23) and (10.4.24) we see that (10.4.22) holds.

Hence, for a fixed x a properly standardized version of $f_n(x)$, has a limiting standard normal distribution.

10.4.1   *Optimal Bandwidth*

Kernel type density estimators define a class of density functions. Let us look at the optimal bandwidth and the optimal kernels in this class of kernels satisfying the following conditions:

(i) $K(.)$ is a bounded density function,
(ii) $K(x) = K(-x) \; \forall x$,
(iii) $\int_{-\infty}^{\infty} x^2 K(x)dx = c_0$,
(iv) the density function $f(x)$ is bounded and twice continuously differentiable with $f, f^{(2)} \in L^2(R)$, the class of square integrable real valued functions.

Using Taylor's series expansion and above conditions $E[f_n(x)]$ can be written as follows

$$E[f_n(x)] = E[\frac{1}{h_n}K(\frac{x - X_i}{h_n})]$$

$$= \int_{-\infty}^{\infty} f(x - zh_n)K(z)dz$$

$$\simeq \int_{-\infty}^{\infty} K(z)[f(x) - zh_n f^{(1)}(x) + \frac{z^2 h_n^2}{2}f^{(2)}(x)]dz$$

$$= f(x) + \frac{h_n^2}{2}f^{(2)}(x). \qquad (10.4.25)$$

Thus, the bias and the variance of $f_n(x)$ can be approximated by

$$E[f_n(x)] - f(x) \simeq \frac{h_n^2}{2} f^{(2)}(x), \qquad (10.4.26)$$

$$Var[f_n(x)] \simeq \frac{1}{nh_n} f(x) \int_{-\infty}^{\infty} K^2(y) dy. \qquad (10.4.27)$$

Therefore, the Mean Square Error of $f_n(x)$ is

$$E[f_n(x) - f(x)]^2 = Var[f_n(x)] + [E[f_n(x)] - f(x)]^2$$

$$\simeq \frac{1}{nh_n} f(x) \int_{-\infty}^{\infty} K^2(y) dy + \frac{1}{4} c_0^2 [f^{(2)}(x) h_n^2]^2.$$

$$(10.4.28)$$

The Mean Integrated Square Error (MISE) of $f_n(x)$ is

$$\int_{-\infty}^{\infty} E[f_n(x) - f(x)]^2 dx = \int_{-\infty}^{\infty} Var[f_n(x)] dx + \int_{-\infty}^{\infty} [E[f_n(x)] - f(x)]^2 dx$$

$$\simeq \frac{1}{nh_n} \int_{-\infty}^{\infty} K^2(y) dy + \frac{1}{4} h_n^4 c_0^2 \int_{-\infty}^{\infty} f^{(2)}(x) dx.$$

$$(10.4.29)$$

Then, the optimal value of the bandwidth h_n which minimizes the MISE in (10.4.29) is given by

$$h_n(opt) = [c_0]^{-\frac{2}{5}} [\int_{-\infty}^{\infty} K^2(y) dy]^{\frac{1}{5}} [\int_{-\infty}^{\infty} f^{(2)}(x) dx]^{-\frac{1}{5}} n^{-\frac{1}{5}}. \qquad (10.4.30)$$

Comments:

(i) The optimal value of the bandwidth h_n given in (10.4.30) depends on the unknown second derivative $f^{(2)}(x)$.

(ii) $\int_{-\infty}^{\infty} f^{(2)}(x) dx$ measures the fluctuations in the unknown density function. If the variations in the density function are large, then one should choose smaller values of h_n.

(iii) Clearly, the optimal choice of h_n converges to 0 as the sample size increases. But the rate of convergence is very slow.

The obvious question then is how does one make an appropriate choice of h_n. One way out is to choose h_n with respect to a standard family of distributions. If $f(x)$ is taken as the the $N(0, \sigma^2)$ density function, then

$$\int_{-\infty}^{\infty} f^{(2)}(x) dx \simeq 0.212 \sigma^{-5},$$

$$h_n(opt) \simeq 1.06 \sigma n^{-\frac{1}{5}}. \qquad (10.4.31)$$

The unknown variance can be estimated from the data. This procedure is subjective, it works well only if the unknown distribution is close to the normal density. In other cases it could oversmooth or undersmooth the density function.

The other way out is the cross-validation procedure for finding the bandwidth which is totally dependent on the observed value of the random variables. The Integrated Square Error of $f_n(x)$ is given by

$$\int_{-\infty}^{\infty} [f_n(x) - f(x)]^2 dx = \int_{-\infty}^{\infty} f_n^2(x) dx - 2 \int_{-\infty}^{\infty} f_n(x) f(x) dx + \int_{-\infty}^{\infty} f^2(x) dx. \tag{10.4.32}$$

Since $f(x)$ is independent of the data, minimizing (10.4.32) with respect to h_n is equivalent to minimizing

$$A(f_n) = \int_{-\infty}^{\infty} f_n^2(x) dx - 2 \int_{-\infty}^{\infty} f_n(x) f(x) dx.$$

Let

$$f_n^{-i}(x) = \frac{1}{h_n(n-1)} \sum_{1 \leq j \neq i \leq n} K\left(\frac{x - X_j}{h_n}\right). \tag{10.4.33}$$

Thus, $f_n^{-i}(x)$ is an estimator of the density function $f(x)$ based on all n observations $X_1, X_2, \ldots, X_n$ except X_i.

Then, an estimator of $A(f_n)$ is given by

$$\frac{1}{n^2 h_n} \sum_{i=1}^{n} \sum_{j=1}^{n} \tilde{K}\left(\frac{X_i - X_j}{h_n}\right) - \frac{1}{(n-1)h_n} K(0), \tag{10.4.34}$$

where

$$\tilde{K} = K^{(2)}(t) - 2K(t), \tag{10.4.35}$$

and $K^{(2)}(t)$ is the convolution of the kernel density $K(x)$ with itself.

10.4.2   *Optimal Kernels*

The approximate value of MISE given in (10.4.29) with optimal choice of h_n (10.4.30) is given by

$$\frac{5}{4} B(K)\left[\int_{-\infty}^{\infty} (f^{(2)}(x))^2 dx\right]^{\frac{1}{5}} n^{-\frac{4}{5}}, \tag{10.4.36}$$

$$B(K) = c_0^{\frac{2}{5}}\left[\int_{-\infty}^{\infty} K^2(t) dt\right]^{\frac{4}{5}},$$

where $B(K)$ is the part of the optimal MISE which depends on the kernel density function and

$$c_0 = \int_{-\infty}^{\infty} t^2 K(t) dt$$

is the second moment of the density function.

Given the correct bandwidth, a smaller value of $B(K)$ gives a smaller value of MISE. That is, one has to minimize $B(K)$ subject to the fact that $K(t)$ is a density function and the second moment $c_0 = 1$. The optimal Kernel subject to these constraints is the Epanechnikov kernel given by

$$K_{EPC}(x) = \begin{cases} \frac{3}{4\sqrt{5}}(1 - \frac{1}{5}x^2) & \text{if} \quad -\sqrt{5} \le x \le \sqrt{5}, \\ 0 & \text{otherwise.} \end{cases}$$

Two different kernels can be compared by looking at their relative MISE. This essentially means comparing corresponding values of $B(K)$ functions. Hence, the efficiency of any symmetric kernel $K(x)$ relative to the optimal kernel $K_{EPC}(x)$ is defined as follows

$$eff(K) = [\frac{B(K_{EPC})}{B(K)}]^{\frac{5}{4}}$$

$$= \frac{3}{5\sqrt{5}}[\int_{-\infty}^{\infty} t^2 K(t) dt]^{-\frac{1}{2}}[\int_{-\infty}^{\infty} K^2(t) dt]^{-1}. \quad (10.4.37)$$

Thus, efficiency of the kernels mentioned earlier with respect to Epanechnikov kernel are given as below

Biweight	Triangular	Gaussian	Rectangular
0.994	0.986	0.951	0.925

The efficiencies of all the kernels above with respect to the optimal kernel (Epanechnikov) are close to 1. This shows that the choice of the kernel is not that important in the constructions of the estimator $f_n(x)$.

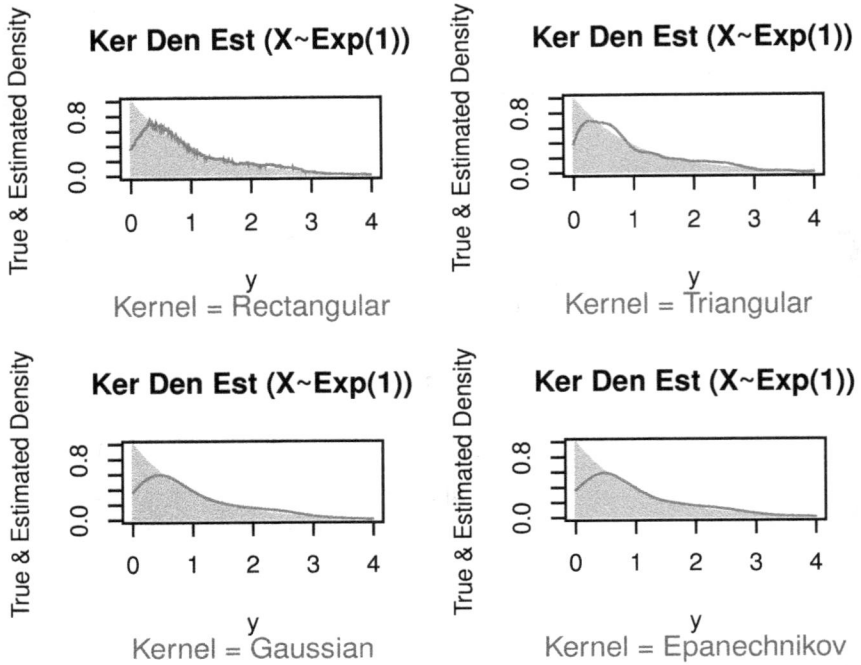

Fig. 10.14   Exact and estimated standard exponential density function with bandwidth 0.3

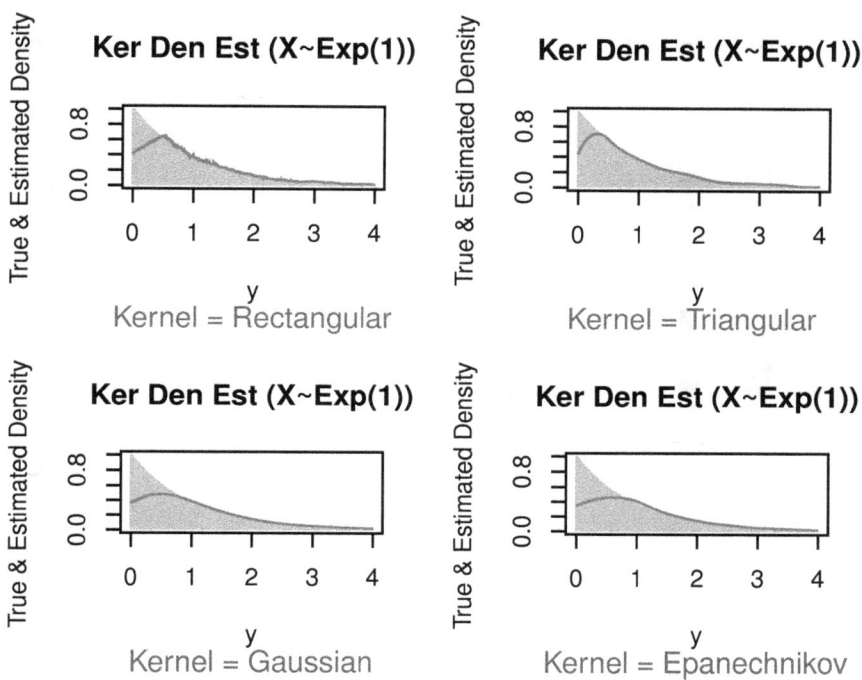

Ker Den Est (X~Exp(1))
Kernel = Rectangular

Ker Den Est (X~Exp(1))
Kernel = Triangular

Ker Den Est (X~Exp(1))
Kernel = Gaussian

Ker Den Est (X~Exp(1))
Kernel = Epanechnikov

Fig. 10.15   Exact and estimated standard exponential density function with bandwidth 0.5

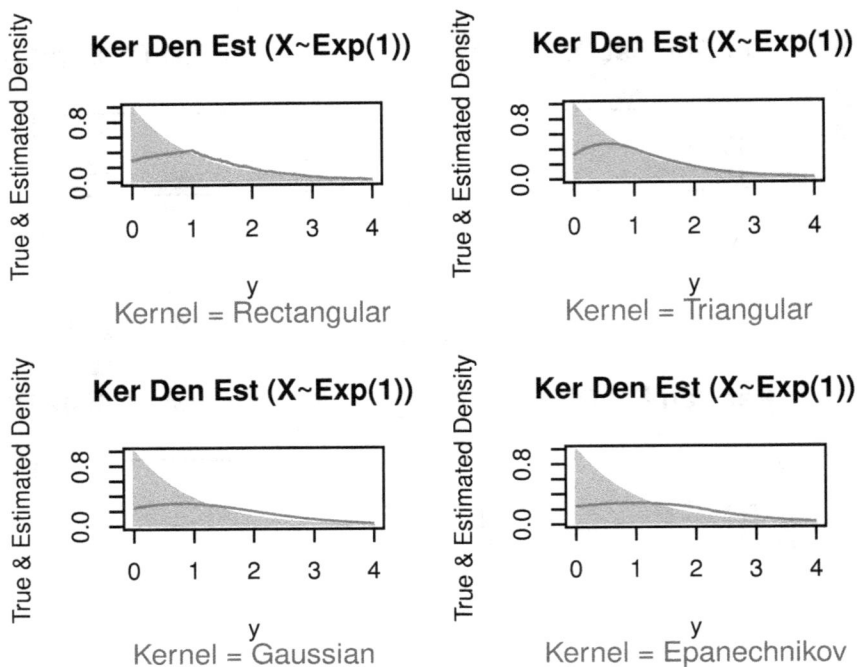

Fig. 10.16 Exact and estimated standard exponential density function with bandwidth 1.0

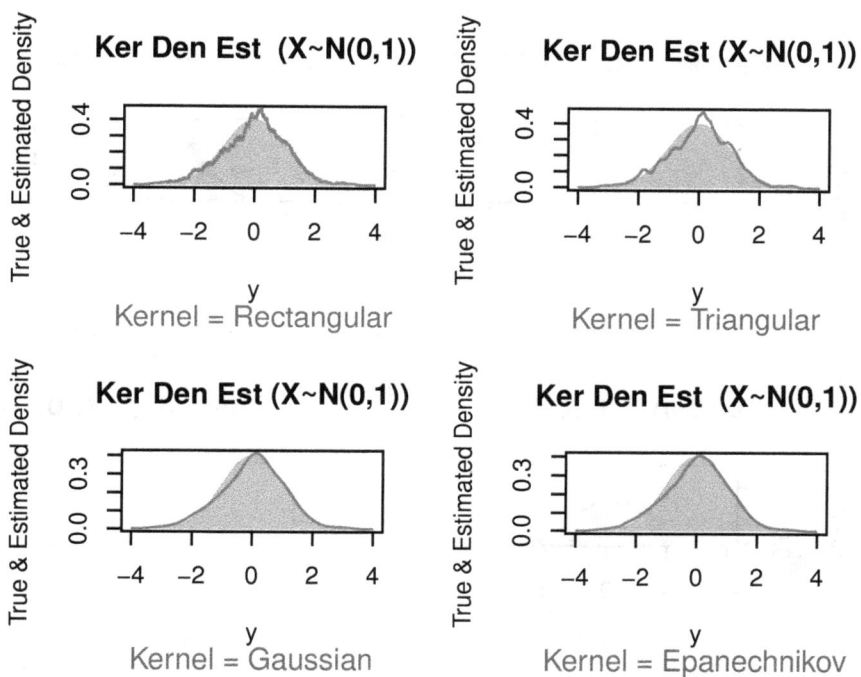

Fig. 10.17 Exact and estimated standard exponential density function with bandwidth 0.3

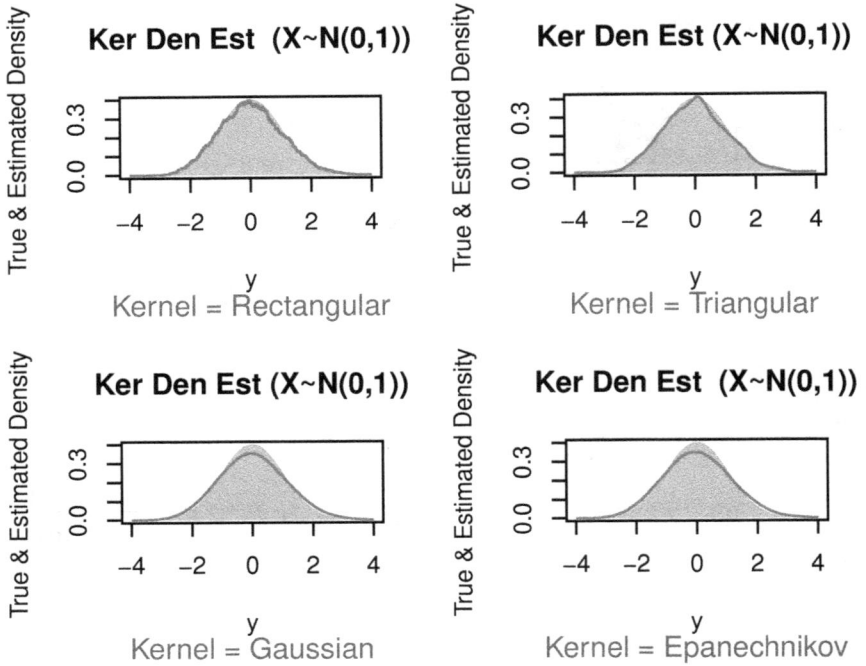

Fig. 10.18   Exact and estimated standard exponential density function with bandwidth 0.5

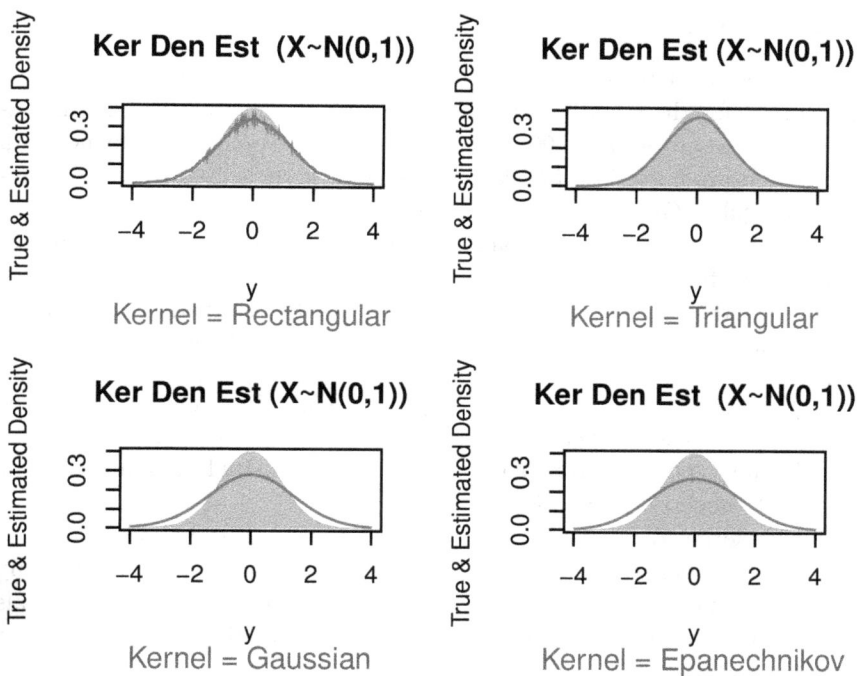

Fig. 10.19   Exact and estimated standard exponential density function with bandwidth 1.0

Comments:

Figures 10.14-10.16 give the actual density function and the kernel type estimators for the standard exponential density function. The density estimator is based on $n = 500$ observations and three choices of bandwidth considered are $h_n = 03, 0.5, 1.0$. Four different kernels used are - rectangular, triangular, Gaussian and Epanechnikov.

Figures 10.17-10.19 give analogous representations for the standard normal density function.

Clearly the choice of kernel is not important. But the kernel type density estimator becomes smoother with the increase in the bandwidth.

(i) The kernel type estimator $f_n(x)$ can be suitably modified when the observations $X_1, X_2, \ldots, X_n$ are subject to random censoring.

(ii) Kernel type estimators of derivatives of the density function (assuming that they exist), can also be found.

(iii) The kernel K is a symmetric density function. If the observations are i.i.d random variables with the density function having support only on the positive half of the real line (as in reliability where one deals with lifetimes), the density estimator $f_n(x)$ would also have support for $x < 0$.

(iv) The failure rate of a positive valued random variable is defined as

$$r(x) = \frac{f(x)}{1 - F(x)}, \quad x > 0.$$

Kernel type estimators of $r(x)$ that involve $f_n(x)$ have been studied both when observations are uncensored and censored.

(v) One can find kernel type estimators for multivariate density functions also.

10.5   Goodness-of-fit Tests Based on Hellinger Distance

Goodness-of-fit tests based on distance between two distribution functions has been discussed in Chapter 4. Such tests may be based on distances or disparities between the probability density functions (p.d.f.). The squared Hellinger distance $HD(f, g)$ between two p.d.f.s f and g is defined as

$$HD(f, g) = \int (f^{1/2}(x) - g^{1/2}(x))^2 dx.$$

(In this section the integration is over the entire range of the variable.)

[Beran (1977)] has suggested a density based approach for testing goodness of fit, which involves estimating the completely unknown p.d.f. of the probability distribution from which the data has been collected using the techniques suggested earlier in the chapter and also simultaneously estimating the parameters of the family of distributions with which the goodness of fit is being tested. This is achieved by deriving a statistic which is the minimum of the Hellinger distance between the estimated density in general and the density estimated under the null hypothesis. The calculation of the statistic is not easy and requires computer intensive approach.

Let $X_1, X_2, \cdots, X_n$ be a random sample from the p.d.f. g and let $\mathcal{F} = \{f_\theta, \theta \in \Theta\}$ be a specified parametric family of p.d.f.s. The aim is to test the hypothesis

$$H_0 : g \in \mathcal{F} \ \ versus \ \ g \notin \mathcal{F}.$$

A minimum Hellinger distance estimator (MHDE) $\hat{\theta}_n$ of θ is the value that minimizes $HD(f_\theta, \hat{g}_n)$, where $\hat{g}_n$ is some nonparametric density estimator of g. That is

$\hat{\theta}_n = $ minimiser $HD(f_\theta, \hat{g}_n) = T(\hat{g}_n)$.

The minimized distance $HD(f_{\hat{\theta}_n}, \hat{g}_n)$ then provides a natural goodness-of-fit statistic.

Asymptotic properties of $\hat{\theta}_n$ and $HD(f_{\hat{\theta}_n}, \hat{g}_n)$ are obtained by [Beran (1977)] for continuous models when $\hat{g}_n$ is a kernel density estimator and under many regularity conditions on the kernel and the densities admitted in the parametric family and when the parameter space Θ is a compact subset a p-dimensional Euclidean space R^p. These conditions are given in the Appendix.

Let

$$\hat{g}_n(x) = \frac{1}{h_n} \sum_{i=1}^{n} K\left(\frac{x - X_i}{h_n}\right)$$

and let the bandwidth $h_n = C_n S_n$ is, where $\{C_n\}$ is a sequence of positive constants, $S_n = S_n(X_1, X_2, \cdots, X_n)$ is a robust scale estimator and the kernel $K(\cdot)$ is a density function.

Let

$$R_n = \max_{1 \le i \le n} X_i - \min_{1 \le i \le n} X_i,$$

$$\mu_n = \frac{1}{4} R_n \int K^2(x) dx$$

and

$$\sigma_n^2 = \frac{1}{8} C_n R_n \int (K * K)^2(x) dx$$

where

$$K * K(x) = \int K(x - t) K(t) dt.$$

Then under the regularity conditions,

$$H_n = \sigma_n^{-1}[n C_n H D(f_{\hat{\theta}_n}, \hat{g}_n) - \mu_n]$$

converges in distribution to a standard normal variable $N(0, 1)$ under f_θ as $n \to \infty$.

Thus the α-level test is to reject H_0 if $|H_n| > z_{1-\alpha}$, where $z_{1-\alpha}$ is the upper $\alpha\%$ value of the standard normal distribution.

Also, the limiting distribution of $\sqrt{n}(\hat{\theta}_n - \theta)$ is normal with mean 0 and variance $\frac{1}{4}[\int \dot{h}_\theta(x) \dot{h}_\theta^T(x) dx]^{-1}$ under f_θ, where $\dot{h}_\theta(x) = (\dot{h}_\theta^{(1)}(x), \cdots, \dot{h}_\theta^{(p)}(x))^T$ with T denoting the transpose.

Comment:

The result holds for parametric families $\{f_\theta, \theta \in \Theta\}$ where Θ is not compact but can be embedded within a compact set. This, for example, is possible for a location scale family $\{\sigma^{-1} f(\sigma^{-1}(x - \mu)); \sigma > 0, -\infty < \mu < \infty\}$ where f is continuous by the transformation $\mu = \tan \theta_1$, $\sigma = \tan \theta_2$

$$\theta = (\theta_1, \theta_2) \in (-\pi/2, \pi/2) X(0, \pi/2) = \Theta'$$

$$f_\theta(x) = (\tan \theta_2)^{-1}[f((\tan \theta_2)^{-1}(x - \tan \theta_1))],$$

$\theta \in \Theta'$ and $f_\theta^{1/2} - g_t^{1/2}$ can be extended to a continuous function on $\overline{\Theta} = [-\pi/2, \pi/2] \times [0, \pi_2]$ which is compact.

As mentioned earlier the calculation of the statistic requires numerical procedures. The package 'mhde' in the software 'R' computes the statistic for testing for normality. We use this package for the example given below.

Example 10.1: A random sample of size 50 was drawn from a standard normal distribution. The 50 realized sample values were:

$$
x_i : \begin{array}{ccccc}
-1.78591244 & -1.84806439 & 0.43908867 & 1.89931392 & 0.11460372 \\
-0.99975553 & 1.98227882 & 1.27408448 & -0.65982870 & -0.17998088 \\
-0.96214672 & -2.77070720 & 1.03211597 & 0.63404868 & -0.01136362 \\
0.17044201 & 0.25887653 & 0.40709031 & -1.25989876 & 0.01786809 \\
-0.99163883 & 0.68777496 & -2.25124910 & -0.92807826 & 0.51053525 \\
-0.06455210 & -0.67074125 & 1.01480298 & -0.67973552 & -0.80499295 \\
0.75452452 & -1.17989824 & -0.50269120 & 0.50929114 & -0.03872944 \\
0.42513314 & -0.31860478 & 0.15257267 & -1.97994943 & 0.94175868 \\
0.40813190 & 1.49367941 & 0.16790680 & 0.19326649 & 1.86317466 \\
-0.75527608 & -1.81606802 & -1.84208663 & 0.52984966 & 0.24812656
\end{array}
$$

The aim is to test $H_0 : f_\theta$ belongs to the family $\{N(\mu, \sigma^2), -\infty < \mu < \infty, \ 0 < \sigma^2 < \infty\}$. The package obtains the MHDEs of μ and σ^2 by an iterative algorithm with the initial estimates as $\hat{\mu}^{(0)} = $ median $\{x_i\}$ and

$$
\hat{\sigma}^{(0)} = (0.6745)^{-1} \text{ median } \{|x_i - \hat{\mu}^{(0)}|\},
$$

where 0.6745 is the 0.75 quantile of the standard normal distribution. Note that $\hat{\sigma}^{(0)}$ is a consistent estimator of the standard deviation of a normal population.

The package uses the density estimator $\hat{g}_n(x)$ based on the Epanechnikov kernel:

$$
K(x) = .75(1 - x^2) \text{ for } |x| \le 1.
$$

For this kernel $\int_{-1}^{1} K(x)^2 dx = 3/5$ and $\int_{-1}^{1} (K * K(x))^2 dx = \frac{167}{355}$. The scale statisic $S_n = \hat{\sigma}^{(0)}$. The value of C_n for sample size $n = 50$ is taken to be 0.7372644.

For the data in the example, $\hat{\mu}^{(0)} = 0.0662359$ and $\hat{\sigma}^{(0)} = 1.08455$. The output of the package gives the final estimates of μ and σ, the Hellinger distance and the p-value. For the above data, these values are: location estimate $= 0.1501589$; Scale estimate $= 1.119368$; Hellinger Distance $= 0.08626999$; p-value $= 0.5434091$. Since the p-value is > 0.1, there is no evidence against the normal distribution.

Comments:

(i) For the discrete models, goodness-of-fit tests based on power divergence statistics have been introduced by [Cressie and Read (1984)] and [Read and Cressie (2012)]. The power divergence I^λ between densities f and g is defined by

$$
I^\lambda(g, f) = \frac{1}{\lambda(\lambda + 1)} \int g(x) \left[\left(\frac{g(x)}{f(x)} \right)^\lambda - 1 \right] dx.
$$

The power divergence statistics of [Cressie and Read (1984)] is of the form

$$I_n^\lambda = \frac{2}{2n\lambda(\lambda+1)} \sum_{i=1}^{k} O_i \left\{ \left(\frac{O_i}{np_i}\right)^\lambda - 1 \right\}, \ \lambda \in R$$

where O_i are the observed frequencies and np_i the expected frequencies. The Pearson's $\chi^2(\lambda = 1)$, log likelihood ratio statistic ($\lambda \to 0$), Freeman-Tukey statistic ($\lambda = -\frac{1}{z}$) are all special cases of the above. The statistic for $\lambda = 2/3$ is shown to be a good alternative to the χ^2 test.

(ii) For the discrete models, goodness-of-fit tests based on the blended weight Hellinger distance methods have been introduced and their comparisons given in [Basu and Sarkar (1994)] and [Shin *et al.* (1995)].

Chapter 11

REGRESSION ANALYSIS

11.1 Introduction

Traditionally regression has been the statistical technique which enables one to predict the values of a random variable Y when in practice another variable is actually observed. This is useful in situations where it is relatively difficult, expensive or time consuming to observe Y than observing X. It is assumed that a training data set in the form of a bivariate random sample $(Y_1, X_1), (Y_2, X_2), \ldots, (Y_n, X_n)$ is available. It is used to devise the prediction rule to be used to predict a new Y value when the corresponding X value is available. Again, assuming a bivariate p.d.f. $f(x, y)$ for X and Y which has moments up to order 2, it is well known that $g(x) = E(Y|X = x)$, the conditional expectation of Y given $X = x$, minimizes the mean squared error $E(Y - g(X))^2$ among all functions. Hence an estimator of $E(Y|x)$ is often used as the predictor of Y. In case (X, Y) have the bivariate normal distribution $BVN(\mu_x, \mu_y, \sigma_x^2, \sigma_y^2, \rho)$, then $E(Y|x)$ is given as

$$E(Y|x) = \mu_Y + \rho \frac{\sigma_Y}{\sigma_X}(x - \mu_x)$$

$$= (\mu_Y - \rho \frac{\sigma_Y}{\sigma_X}\mu_x) + \rho \frac{\sigma_Y}{\sigma_X}x$$

$$= \beta_0 + \beta_1 x,$$

where

$$\beta_1 = \rho \frac{\sigma_Y}{\sigma_X}, \quad \beta_0 = \mu_Y - \beta_1 \mu_X.$$

Hence, in this case $E(Y|x)$ is a linear function of x and the problem of estimating the conditional expectation reduces to estimating the constants β_0 and β_1. The estimators are given by

$$\hat{\beta}_1 = \hat{\rho} \frac{\hat{\sigma}_Y}{\hat{\sigma}_X}, \quad \hat{\beta}_0 = \bar{y} - \hat{\beta}_1 \bar{x},$$

189

where

$$\bar{x} = \frac{1}{n} \sum_{i=1}^{n} x_i, \quad \bar{y} = \frac{1}{n} \sum_{i=1}^{n} y_i,$$

$$\hat{\sigma}_X^2 = \frac{1}{n} \sum_{i=1}^{n} (x_i - \bar{x})^2, \quad \hat{\sigma}_Y^2 = \frac{1}{n} \sum_{i=1}^{n} (y_i - \bar{y})^2,$$

$$\hat{\rho} = \frac{\sum_{i=1}^{n} (x_i - \bar{x})(y_i - \bar{y})}{\sqrt{\sum_{i=1}^{n} (x_i - \bar{x})^2 \sum_{i=1}^{n} (y_i - \bar{y})^2}}.$$

In the nonparametric set up we do not assume any functional form for the joint p.d.f. $f(x, y)$. Given two random variables Y and X such that it is suspected that values of X affect the values of Y, the relation may be modelled as

$$Y = g(X) + \epsilon, \tag{11.1.1}$$

where the function $g(x)$ is called the regression of Y on X and ϵ is the random error. The problem is to estimate the regression function $g(x)$. The only assumptions made are that ϵ is a random variable such that

$$E(\epsilon) = 0,$$
$$Var(\epsilon) = \sigma^2 < \infty,$$
$$E(Y|x) = g(x). \tag{11.1.2}$$

So often, the conditional expectation is called the regression of Y on x.

11.2   Least Squares Estimators

In this section we use the least squares estimators to get univariate and the multivariate linear regression models.

11.2.1   *Univariate Regression*

Suppose Y and X are linearly related, where X is the **predictor** variable and Y is the **response** variable. The model

$$Y = \beta_0 + \beta_1 x + \epsilon \tag{11.2.1}$$

involves a single variable X and is called a **simple linear regression** model.

Because of (11.1.2), we have

$$E(Y|x) = \beta_0 + \beta_1 x,$$
$$Var[Y|x] = \sigma^2. \tag{11.2.2}$$

The regression line is

$$g(x) = \beta_0 + \beta_1 x. \tag{11.2.3}$$

The slope β_1 represents the change in mean $E(Y|x)$ for a unit change in x.

We use the method of least squares to estimate β_0 and β_1 on the basis of sample observations $(y_1, x_1), (y_2, x_2), \ldots, (y_n, x_n)$.

Sum of squares of errors is

$$S(\epsilon) = \sum_{i=1}^{n} (y_i - \beta_0 - \beta_1 x_i)^2. \tag{11.2.4}$$

Minimizing (11.2.4) w.r.t. β_0 and β_1 gives the least squares estimators of the intercept β_0 and the slope β_1 as

$$\hat{\beta}_0 = \bar{y} - \hat{\beta}_1 \bar{x},$$
$$\hat{\beta}_1 = \frac{\sum_{i=1}^{n} y_i x_i - \bar{x}\bar{y}}{\sum_{i=1}^{n} x_i^2 - n\bar{x}^2}. \tag{11.2.5}$$

Then, the least squares fitted regression line is given by

$$\hat{y} = \hat{\beta}_0 + \hat{\beta}_1 x. \tag{11.2.6}$$

Then the ith residual is given by

$$e_i = y_i - \hat{y}_i$$
$$= y_i - (\hat{\beta}_0 + \hat{\beta}_1 x_i), \quad i = 1, 2, \ldots, n. \tag{11.2.7}$$

The *ith* residual represents the difference between the observed value y_i and the fitted value $\hat{y}_i$. Note that

$$\hat{\beta}_1 = \sum_{i=1}^{n} c_i y_i,$$

where $c_i = \frac{x_i - \bar{x}}{S_{xx}}$, $S_{xx} = \sum_{i=1}^{n} (x_i - \bar{x})^2$.

Thus, $\hat{\beta}_0$ and $\hat{\beta}_1$ can be expressed as linear combinations of observed values $y_1, y_2, \ldots, y_n$. It is easy to see that

$$E(\hat{\beta}_0) = \beta_0, \quad E(\hat{\beta}_1) = \beta_1.$$

Hence both $\hat{\beta}_0$ and $\hat{\beta}_1$ are unbiased estimators of β_0 and β_1, respectively. Since $Y_1, Y_2, \ldots, Y_n$ are independent and $Cov(\bar{Y}, \hat{\beta}_1) = 0$, we have

$$Var(\hat{\beta}_1) = \frac{\sigma^2}{S_{xx}}, \quad Var(\hat{\beta}_0) = \sigma^2 \left(\frac{1}{n} + \frac{\bar{x}^2}{S_{xx}} \right).$$

From the Gauss-Markov Theorem, it follows that the least squares estimators $\hat{\beta}_0$ and $\hat{\beta}_1$ have minimum variance in the class of all unbiased estimators of β_0 and β_1 that are linear combinations of y_i. Hence the estimators are known to be the Best Linear Unbiased estimators (BLUE).

Comments:

(i) $\sum_{i=1}^{n} e_i = \sum_{i=1}^{n} (y_i - \hat{y}_i) = 0$. That is, the sum of residuals in a linear regression model is zero. Equivalently, $\sum_{i=1}^{n} y_i = \sum_{i=1}^{n} \hat{y}_i$. That is, the sum of observed y_i's is the same as the sum of expected y_i's.

(ii) The regression line (11.2.6) passes through $(\bar{y}, \bar{x})$.

(iii) The sum of squares of residuals is given by

$$\sum_{i=1}^{n} e_i^2 = \sum_{i=1}^{n} (y_i - \hat{y}_i)^2$$

$$= \sum_{i=1}^{n} (y_i - \bar{y})^2 - \hat{\beta}_1 S_{xy},$$

where $S_{xy} = \sum_{i=1}^{n} y_i (x_i - \bar{x})$.

Therefore, an unbiased estimator of σ^2 is given by

$$\hat{\sigma}^2 = \frac{\sum_{i=1}^{n} e_i^2}{n-2}.$$

11.2.2 *Multiple Linear Regression*

Suppose the response variable Y depends on p regressor variables $X_1, X_2, \ldots, X_p$. Again assuming a linear regression model, one can write

$$\mathbf{Y} = \beta \mathbf{X} + \epsilon, \tag{11.2.8}$$

where $\mathbf{Y}$ is the n-vector of observations on Y, $\mathbf{X}$ is the $n \times p$ matrix of n observations on the p-variate predictor variable and the elements of ϵ, are the random and uncorrelated errors with mean $\mathbf{0}$ and common variance σ^2. The least squares estimator of the p-vector β of coefficients is given by

$$\hat{\beta} = (\mathbf{X}'\mathbf{X})^{-1}\mathbf{X}'\mathbf{Y} = \mathbf{HY}, \tag{11.2.9}$$

which are linear functions of elements of $\mathbf{Y}$. The matrix $\mathbf{H} = (\mathbf{X}'\mathbf{X})^{-1}\mathbf{X}'$ is based on elements of $\mathbf{X}$ alone and is called the **hat matrix**.

11.3   Shrinkage Estimators

Sometimes there is a belief that the standard estimators of regression parameters should be modified so that they tilt towards a known value of the parameter (without loss of generality, say θ). This can be achieved through the regression techniques - Ridge and Least Absolute Selection and Shrinkage Operator estimators described in this section. These estimators are not unbiased but seen to possess a smaller M.S.E. compared to the best unbiased estimators.

11.3.1   *Ridge Estimators*

Consider the multiple linear regression problem. Suppose that β^* is a biased estimator of β. Then the M.S.E. of β^* is seen to be

$$E[\beta^* - \beta]^2 = Var[\beta^*] + [E(\beta^*) - \beta]^2. \qquad (11.3.1)$$

Thus, the M.S.E. of β^* is the sum of the variance of the biased estimator and the square of the bias of the estimator.

The least squares estimator $\hat{\beta}$ is the BLUE. Hence the bias is zero. However, it is possible that the variance of the estimator is large and one could use a biased estimator of β whose variance is smaller than that of the unbiased estimator $\hat{\beta}$ and even its M.S.E. is smaller than that of the unbiased estimator.

Besides it is possible that some components $\hat{\beta}_j$, $1 \le j \le p$ of the vector $\hat{\beta}$ take very small values. The corresponding variables X_j may not be contributing significantly to the regression model.

In the multiple linear regression set up one could look at another class of estimators, the *ridge* estimators which are biased, whose variance is smaller than that of the unbiased estimator $\hat{\beta}$ and the $\hat{\beta}'_j s$ that are too small are shrinking towards zero.

Ridge estimator $\hat{\beta}_R$ can be found by minimizing

$$\sum_{i=1}^{n}(y_i - \beta_0 - \sum_{j=1}^{p}\beta_j x_{ij})^2 + \lambda \sum_{j=1}^{p}\beta_j^2, \qquad (11.3.2)$$

over $\lambda \ge 0$.

The second term $\sum_{j=1}^{p}\beta_j^2$ in (11.3.2) is called the *penalty* term. The parameter β_0 has been left out of the penalty term as we want the solution to be independent of the origin. Since $\hat{\beta}_0 = \bar{y}$ we shift x_{ij} to $x_{ij} - \bar{x}$ and minimize (11.3.2) as if there was no intercept.

The parameter λ above is called the *tuning* parameter. Note that when $\lambda = 0$, we get the usual linear regression. When $\lambda = \infty$, the ridge estimator $\hat{\beta}_R = \mathbf{0}$. The tuning parameter λ is chosen so as to reduce the prediction error given in (11.3.2). In practice, we calculate the M.S.E.(λ) for some values of λ, $0 < \lambda < \infty$ and choose the one which seems to reduce the M.S.E.(λ) as much as possible.

The residual sum of squares are given by

$$RSS = (\mathbf{Y} - \mathbf{X}\beta)'(\mathbf{Y} - \mathbf{X}\beta) + \lambda\beta'\beta. \qquad (11.3.3)$$

Differentiation w.r.t β, we get $\hat{\beta}_R$ is a solution of the equation

$$(\mathbf{X}'\mathbf{X} + \lambda\mathbf{I})\hat{\beta}_R = \mathbf{X}'\mathbf{Y}. \qquad (11.3.4)$$

Hence, the ridge estimator $\hat{\beta}_R$ is given by

$$\hat{\beta}_R = [(\mathbf{X}'\mathbf{X} + \lambda\mathbf{I})]^{-1}\mathbf{X}'\mathbf{Y}. \qquad (11.3.5)$$

Note that $(\mathbf{X}'\mathbf{X} + \lambda\mathbf{I})$ will be non-singular even if $\mathbf{X}'\mathbf{X}$ is singular. Using (11.2.9), we get

$$\hat{\beta}_R = [(\mathbf{X}'\mathbf{X} + \lambda\mathbf{I})]^{-1}\mathbf{X}'\mathbf{X}\hat{\beta}$$
$$= \mathbf{C}_\lambda\hat{\beta}, \qquad (11.3.6)$$

where $\mathbf{C}_\lambda = (\mathbf{X}'\mathbf{X} + \lambda\mathbf{I})]^{-1}\mathbf{X}'\mathbf{X}$. Hence, $\hat{\beta}_R$ is a linear function of elements of $\hat{\beta}$.

Since $\hat{\beta}$ is an unbiased estimator of β, we have

$$E(\hat{\beta}_R) = \mathbf{C}_\lambda\beta. \qquad (11.3.7)$$

Hence $\hat{\beta}_R$ is a biased estimator of β. Variance-covariance matrix of $\hat{\beta}_R$ is given by

$$Var[\hat{\beta}_R] = \sigma^2[(\mathbf{X}'\mathbf{X} + \lambda\mathbf{I})]^{-1}\mathbf{X}'\mathbf{X}[(\mathbf{X}'\mathbf{X} + \lambda\mathbf{I})]^{-1}. \qquad (11.3.8)$$

The M.S.E. of the ridge estimator $\hat{\beta}_R$ is

$$M.S.E.(\hat{\beta}_R) = Var[\hat{\beta}_R] + (Bias[\hat{\beta}_R])^2$$
$$= \sigma^2[(\mathbf{X}'\mathbf{X} + \lambda\mathbf{I})]^{-1}\mathbf{X}'\mathbf{X}[(\mathbf{X}'\mathbf{X} + \lambda\mathbf{I})]^{-1} + \lambda^2\beta'[(\mathbf{X}'\mathbf{X} + \lambda\mathbf{I})]^{-2}\beta$$
$$= \sigma^2 tr[[(\mathbf{X}'\mathbf{X} + \lambda\mathbf{I})]^{-1}\mathbf{X}'\mathbf{X}[(\mathbf{X}'\mathbf{X} + \lambda\mathbf{I})]^{-1}] + \lambda^2\beta'[(\mathbf{X}'\mathbf{X} + \lambda\mathbf{I})]^{-2}\beta$$
$$= \sigma^2 \sum_{j=1}^{p} \frac{c_i}{(c_i + \lambda)^2} + \lambda^2\beta'[(\mathbf{X}'\mathbf{X} + \lambda\mathbf{I})]^{-2}\beta, \qquad (11.3.9)$$

where $c_i's$ are eigen values of $\mathbf{X}'\mathbf{X}$.

Note that as λ increases, the bias increases and the variance decreases. We choose λ such that the reduction in variance is greater than increase in squared bias.

Further the residual sum of squares of $\hat{\beta}_R$ can be expressed as follows

$$(\mathbf{Y} - \mathbf{X}\hat{\beta}_R)'(\mathbf{Y} - \mathbf{X}\hat{\beta}_R)$$
$$= (\mathbf{Y} - \mathbf{X}\hat{\beta})'(\mathbf{Y} - \mathbf{X}\hat{\beta}) + (\hat{\beta}_R - \hat{\beta})'\mathbf{X}'\mathbf{X}(\hat{\beta}_R - \hat{\beta}).$$

$$(11.3.10)$$

Hence the residual sum of squares for $\hat{\beta}_R$ is the residual sum of squares of $\hat{\beta}$ plus a term giving the square of scaled bias.

[Hoerl *et al.* (1975)] suggested that we chose λ as

$$\lambda = \frac{k\hat{\sigma}^2}{\hat{\beta}'\hat{\beta}},$$

where $\hat{\beta}$ and $\hat{\sigma}^2$ are the least squares estimators of β and σ^2.

In practice, as mentioned earlier, M.S.E.(λ) is found for several choices of λ. And the λ which minimizes the M.S.E. is used as the tuning parameter for finding $\hat{\beta}_R$.

Comments:

(i) The problem of finding ridge estimators is equivalent to minimizing $\sum_{i=1}^{n}(y_i - \beta_0 - \sum_{j=1}^{p}\beta_j x_{ij})^2$ subject to the constraint $\sum_{j=1}^{p}\beta_j^2 \le t$. In this case t is the **tuning** parameter. Hence, the ridge estimator $\hat{\beta}_R$ is the set of $\beta_j's$ which minimize the penalized sum of squares subject to constraints.

(ii) Predictor variables need to have the same scale. Else the penalty term $\sum_{j=1}^{p}\beta_j^2$ will not give equal importance to all variables. And the final conclusions could be misleading.

(iii) If all $\beta_j's$ are moderately large, then the advantage of ridge estimators is much smaller than in the case when some $\beta_j's$ are small. In the former case the choice of values of λ which reduce the variance of ridge estimators is usually small.

(iv) Ridge regression is not useful in variable selection directly but can help in deciding which of the regressor variables do not contribute much to the response and may then be considered as candidates for elimination from the study.

11.3.2 *Least Absolute Selection and Shrinkage Operator Estimators*

The ridge estimators can not 'shrink' any of the β_j co-efficients to zero. As an alternative one could use the Least Absolute Selection and Shrinkage Operator (LASSO) estimators introduced in [Tibshirani (1996)]. The

LASSO estimator $\hat{\boldsymbol{\beta}}_L$ is found by minimizing

$$\sum_{i=1}^{n}(y_i - \beta_0 - \sum_{j=1}^{p}\beta_j x_{ij})^2 + \lambda\sum_{j=1}^{p}|\beta_j|, \qquad (11.3.11)$$

over $\lambda \geq 0$.

The motivation for the LASSO estimator is the same as that for the ridge estimator - to have a linear regression for Y on $\boldsymbol{X}$. Comparing (11.3.2) and (11.3.11) one sees that the only difference between the LASSO estimator and the ridge estimator is that the former uses the penalty $\sum_{j=1}^{p}|\beta_j|$ and the latter uses the penalty $\sum_{j=1}^{p}\beta_j^2$. Analogous to the ridge estimator $\hat{\boldsymbol{\beta}}_R$, the LASSO estimator $\hat{\boldsymbol{\beta}}_L$ is same as the least squares estimator $\hat{\boldsymbol{\beta}}$ for $\lambda = 0$ and $\hat{\boldsymbol{\beta}}_L = \boldsymbol{0}$, if $\lambda = \infty$.

Comments:

(i) The problem of finding LASSO estimators is equivalent to minimizing $\sum_{i=1}^{n}(y_i - \beta_0 - \sum_{j=1}^{p}\beta_j x_{ij})^2$ subject to the constraint $\sum_{j=1}^{p}|\beta_j| \leq t$. In this case t is the **tuning** parameter. Hence, the LASSO estimator $\hat{\boldsymbol{\beta}}_L$ is the set of $\beta_j's$ which minimize the penalized sum of squares subject to constraints $\sum_{j=1}^{p}|\beta_j| \leq t$.

(ii) We can not find a closed form expression for the LASSO estimator $\hat{\boldsymbol{\beta}}_L$.

(iii) Value of some coefficients β_j can be equal to 0. Hence the LASSO estimator $\hat{\boldsymbol{\beta}}_L$ is useful in variable selection. The variables corresponding to $\beta_j = 0$ can be dropped from the regression model. Note that, the Ridge and the LASSO regression techniques minimize the residual sum of squares under constraints on the parameters, viz $||\beta||_2 \leq t$ and $||\beta||_1 \leq t$ for some t. The minimizing solution in the $||.||_2$ cannot have any components equal to 0, whereas the one which minimizes the $||.||_1$ may have some 0 components.

(iv) Consider a linear regression model

$$y_i = \mu + \epsilon_i, \quad i = 1, 2, \ldots, n,$$

where

$$E(\epsilon_i) = 0, Var(\epsilon_i) = \sigma^2, Cov(\epsilon_i, \epsilon_j) = 0, \quad i \neq j.$$

Degrees of freedom of the estimate $\hat{y}$ are given by

$$df(\hat{y}) = \frac{1}{\sigma^2}\sum_{i=1}^{n}Cov(\hat{y}_i, y_i).$$

Suppose $\boldsymbol{X}$ is a $n \times p$ matrix.

(a) If $\hat{\beta}$ is a least squares estimator, then
$$df(\hat{y}) = p.$$
(b) If $\hat{\beta}_R$ is the ridge estimator, then
$$df(\hat{y}) = tr(\mathbf{X}[\mathbf{X}'\mathbf{X} + \lambda\mathbf{I}]^{-1}\mathbf{X}').$$
(c) If $\hat{\beta}_L$ is the ridge estimator, then
$$df(\hat{y}) = E[\text{number of nonzero coefficients in } \hat{\beta}_L].$$

(v) Like all multiple regression methods both Ridge and LASSO estimation is heavily computer intensive. The multiple linear regression and ridge regression involves calculation of inverses of matrices which could be of large order. The minimization for LASSO needs to be done numerically. Least Angular Regression (LARS) [Efron *et al.* (2004)] and other time saving algorithms have been developed for this purpose. Most of the popular techniques are included in proprietary softwares such as SPSS, SAS or in open source softwares such as R and are recommended for use in actual practice.

It is obvious that without further distribution assumptions (like normality) one can not get exact variances, confidence intervals or other inference procedures. But asymptotic distributions which exist in far more generality can be used to obtain approximate properties of these procedures. Use of CLT based approximations or those based on bootstrap and simulation techniques and also the so called cross validation techniques may also be used.

Lita da Silva (2013) [Silva (2014)] proved the strong consistency of the ridge estimators using the results for strong consistency of the least squares estimators in multiple regression models under the assumption
$$E(|\epsilon|^r) < \infty \quad \text{for some } r \in (0,2),$$
$$E(\epsilon) = 0, \quad r > 1.$$
[Silva *et al.* (2015)] proved similar results under the assumption that $E(\epsilon) \neq 0$, $E(|\epsilon|^r) < \infty$ for some $r \in (0,1]$.

[Knight and Fu (2000)] considered the asymptotic distribution of the regression estimators when the penalty term is $\sum_{j=1}^{p} |\beta_j|^{\gamma}$ when $\gamma > 0$. For $\gamma = 1$ this would be the LASSO estimator and for $\gamma = 2$ it would be the ridge estimator. [Chatterjee and Lahiri (2011)] showed that if the error terms have finite mean the LASSO estimator is strongly consistent if $\lambda = \lambda_n = o(n)$. However, the consistency fails if $\frac{\lambda_n}{n} \to a$ as $n \to \infty$ $a \in (0,\infty)$. They obtained the rate of convergence to the true parameter if the error variables have a finite moment of order α, $1 < \alpha < 2$.

11.4   Linear Basis Expansion

The regression function $g(\mathbf{X}) = E(Y|\mathbf{X})$ may not be a linear function of $\mathbf{X}$. The linear function holds for the bivariate normal model. It is also easy to understand and interpret and can be visualized as the first order Taylor seies approximation of any function $g(\mathbf{X})$.

This idea can be generalized as follows. Consider the model

$$g(\mathbf{X}) = \sum_{m=1}^{M} \beta_m h_m(\mathbf{X}), \qquad (11.4.1)$$

where $h_m : R^p \to R$ is the mth 'basis function', $m = 1, 2, \ldots, M$.

The basis functions $h_m(\mathbf{X})$ are prespecified transformations of $\mathbf{X}$. The model (11.4.1) is linear in the transformed variables $h_1(\mathbf{X}), h_2(\mathbf{X}), \ldots, h_M(\mathbf{X})$. Hence, the least squares theory can be used to estimate the parameters $\beta_1, \beta_2, \ldots, \beta_M$ and fit the model (11.4.1).

Examples of transformed variables $h_m(\mathbf{X})$ commonly used are

(i) $h_m(\mathbf{X}) = X_m$.
      This gives us the standard version of the linear model (11.2.8).
(ii) $h_m(\mathbf{X}) = X_m^2$ or $h_m(\mathbf{X}) = X_i X_j$.
      This transformation introduces the second order terms, the cross products being interpreted as interactions.
(iii) $h_m(\mathbf{X}) = log X_m$ or $h_m(\mathbf{X}) = \sqrt{X_m}$ or $h_m(\mathbf{X}) = ||\mathbf{X}||$.
      These are the nonlinear transformations used if data suggests such relationships between Y and $\mathbf{X}$.
(iv) $h_m(\mathbf{X}) = I(L_m \leq X_i \leq U_m)$.
      Here $h_m(\mathbf{X})'s$ are indicator functions of a given region.

Choice of polynomial functions could have its disadvantages - as each observation affects the entire curve. This usually results in biased estimators and increase in variance of the estimators at the end of the support of x. So instead of global transformations one could look at local basis functions, so that a given observation affects only the fit close to it and has no effect in fitting the model in other parts of the real line.

11.4.1 *Piecewise Polynomials and Splines*

Let us partition the range of x into 3 disjoint intervals by choosing end points ξ_1, ξ_2. These ξ_1, ξ_2 are called knots. Suppose

$$h_1(x) = I(x \leq \xi_1),$$
$$h_2(x) = I(\xi_1 < x \leq \xi_2),$$
$$h_3(x) = I(x > \xi_2). \tag{11.4.2}$$

In this case the regression function $g(x)$ is given by

$$g(x) = \sum_{m=1}^{3} \beta_m h_m(x),$$
$$= \beta_1 \quad \text{if} \quad x \leq \xi_1,$$
$$\beta_2 \quad \text{if} \quad \xi_1 < x \leq \xi_2,$$
$$\beta_3 \quad \text{if} \quad x > \xi_2. \tag{11.4.3}$$

One can find estimators $\hat{\beta}_1, \hat{\beta}_2, \hat{\beta}_3$ using least squares theory.

Or one could fit a piecewise linear model in each of the sub-regions. That is,

$$g(x) = \beta_1 + \beta_2 x \quad \text{if} \quad x \leq \xi_1,$$
$$\beta_3 + \beta_4 x \quad \text{if} \quad \xi_1 < x \leq \xi_2,$$
$$\beta_5 + \beta_6 x \quad \text{if} \quad x > \xi_2. \tag{11.4.4}$$

However, this procedure gives us a regression function $g(x)$ which is discontinuous at ξ_1 and ξ_2. In order to get a continuous regression function $g(x)$, we must have

$$\beta_1 + \beta_2 \xi_1 = \beta_3 + \beta_4 \xi_1$$
$$\beta_3 + \beta_4 \xi_2 = \beta_5 + \beta_6 \xi_2. \tag{11.4.5}$$

This means we have to find least squares estimators of 6 parameters subject to 2 conditions given in (11.4.5) so as to get a regression function which is continuous at ξ_1, ξ_2. This leads us to the following choice of basis functions

$$h_1(x) = 1,$$
$$h_2(x) = x,$$
$$h_3(x) = (x - \xi_1)_+,$$
$$h_4(x) = (x - \xi_2)_+, \tag{11.4.6}$$

where

$$h_+(x) = \begin{cases} h(x) & \text{if } h(x) \geq 0, \\ 0 & \text{if } h(x) < 0. \end{cases} \qquad (11.4.7)$$

The choice of the basis functions gives us a continuous function $g(x)$ which is linear everywhere except at the end points ξ_1 and ξ_2. The slope of the function in the three regions is different. Note that there are 2 parameters to be estimated in each of the three regions subject to 2 constraints. Hence, there are $3 \times 2 - 2 = 4$ degrees of freedom. Hence we need 4 basis functions to estimate 4 parameters.

In general, one could partition the range of x into $K + 1$ intervals by choosing K knots $\xi_1, \xi_2, \ldots, \xi_K$.

Above is an example of spline. One could generalize it to a piecewise $m - 1$ degree polynomial that is continuous upto $m - 2$ derivatives.

If the number of knots is small and the degree of the polynomial in the region is small, then the basis of the fitted model would be of high dimension. On the other hand increasing the number of knots and degree of the polynomial could result in over fitting and hence result in higher variance.

11.4.2   Cubic Splines

Here we consider fitting a cubic spline. Suppose $\xi_1, \xi_2, \ldots, \xi_K$ are K knots partitioning the range of x into $K + 1$ intervals. The data in each interval is to be fitted by a cubic function. Hence there are $(3 + 1)(K + 1)$ parameters. The continuity of the function $g(x)$ at each of the knots imposes $3K$ constraints. Hence the degrees of freedom are $4(K + 1) - 3K = K + 4$. Hence there are $K + 4$ parameters that have to be estimated. Consider

$$g(x) = \sum_{j=0}^{3} \beta_j x^j + \sum_{\ell=1}^{K} \beta_{3+\ell}(x - \xi_\ell)_+^3. \qquad (11.4.8)$$

The least squares theory is used to estimate $(\beta_0, \beta_1, \ldots, \beta_K)$.

Cubic splines are the most popular type of splines used in practice. Users claim that splines of degree 3 is the lowest order that is needed to ensure continuity at the knots.

11.4.3   Choosing the Knots

Here we have described the fitting of cubic splines. Number of knots can be chosen depending on the shape of the data. These are suitable when there are $1, 2$ or 3 turns in the shape of the data in the given intervals.

Small number of knots will lead to large bias and small variance, whereas large number of knots leads to overfitting: less bias but large variance.

The choice of the knots is in the hand of the user. A natural choice would be to use the quantiles as knots. This would give us intervals with the same number of observations in each one of them. The other possibility is to have knots which are equidistant, that is, we have intervals with the same width.

Smoothing splines use observations as knots. Step-wise regression gives an automatic way of selecting the knots.

11.4.4   *Choosing the Basis Functions*

The choice of the basis functions depends on the properties we are looking for in the regression function. For example, if we use a basis of functions which have continuous first order derivatives, then the regression function will also have continuous first order derivatives. The functions $[(x-\xi_j)_+]^2 = (x - \xi_j)^2_+$ have continuous first order derivatives. Hence, the basis for a quadratic spline function is

$$\{1, x, x^2, (x - \xi_1)^2_+, (x - \xi_2)^2_+, \ldots, (x - \xi_K)^2_+\}. \qquad (11.4.9)$$

This basis helps in identifying the concavity and convexity seen in a scatter plot.

The basis for a power function of order p is

$$\{1, x, x^2, \ldots, x^p, (x - \xi_1)^p_+, (x - \xi_2)^p_+ \ldots, (x - \xi_K)^p_+\}. \qquad (11.4.10)$$

Consider the function $g(x)$ given by

$$g(x) = \sum_{j=0}^{p} \beta_j x^j + \sum_{\ell=1}^{K} \beta_{p+\ell}(x - \xi_\ell)^p_+. \qquad (11.4.11)$$

Then the function $g(x)$ would have continuous derivatives of order $p-1$.

Comments:

(i) Once the knots are fixed the standard least squares theory is used for estimation of the parameters of the model.

(ii) Suppose the fitted model is

$$\hat{\mathbf{y}} = \mathbf{X}\hat{\beta},$$

where $\hat{\boldsymbol{\beta}}$ minimizes $||\hat{y} - X\hat{\beta}||^2$,

$\hat{\beta} = (\beta_0, \beta_1, \ldots, \beta_{K+3})'.$

$$X = \begin{pmatrix} 1 & x_1 & x_1^2 & x_1^3 & (x_1 - \xi_1)_+ & \cdots & (x_1 - \xi_k)_+ \\ \vdots & \vdots & \vdots & \vdots & \vdots & \cdots & \vdots \\ 1 & x_n & x_n^2 & x_n^3 & (x_n - \xi_1)_+ & \cdots & (x_n - \xi_k)_+ \end{pmatrix}$$

One could estimate $(\beta_4, \beta_5, \ldots, \beta_{K+3})'$ subject to the constraint $\sum_{\ell=1}^{K} \beta_{3+\ell}^2 < C$ for a suitable choice of C. (penalized splines)

11.5   An Example

We illustrate the methods proposed in earlier sections on "Concrete comprehensive strength data set" available in the link 'https://archive.ics.uci.edu/ml/datasets/Concrete+Slump+Test'. The data set consists of 103 observations on 7 input variables - Cement, Blast furnace slag (BFS), Fly ash (FA), Water, Superplasticiser (SP), Coarse aggregate (CA) and Fine aggregate. There are 3 output variables - Slump, Flow and 28-day Compressive strength. We use the 'glmnet' function of the glmnet package from 'R' to perform the LASSO and the Ridge regression (for details see [James *et al.* (2013)]).

First we look at LASSO Regression of 'Flow' on the first 6 input variables.

Figure 11.1 gives the Mean Square Error of the LASSO regression plotted against the $\log(\lambda)$. The optimal value of λ in this case is 0.896. The coefficients of the six parameters corresponding to the optimal value of the tuning parameter are given by

Table 11.1 - Regression Parameters under LASSO

Constant	Cement	BFS	FA	Water	SP	CA
-42.893	0	-0.075	0	0.499	0	0

Figure 11.2 shows the Mean Square Error of the Ridge regression plotted against the $\log(\lambda)$. The optimal value of λ in this case is 1.2. The coefficients of the six parameters corresponding to the optimal value of the tuning parameter are given in Table 11.2.

Clearly LASSO regression helps us in variable selection. Clearly Blast furnace Slag and Water seem to effect the flow of cement.

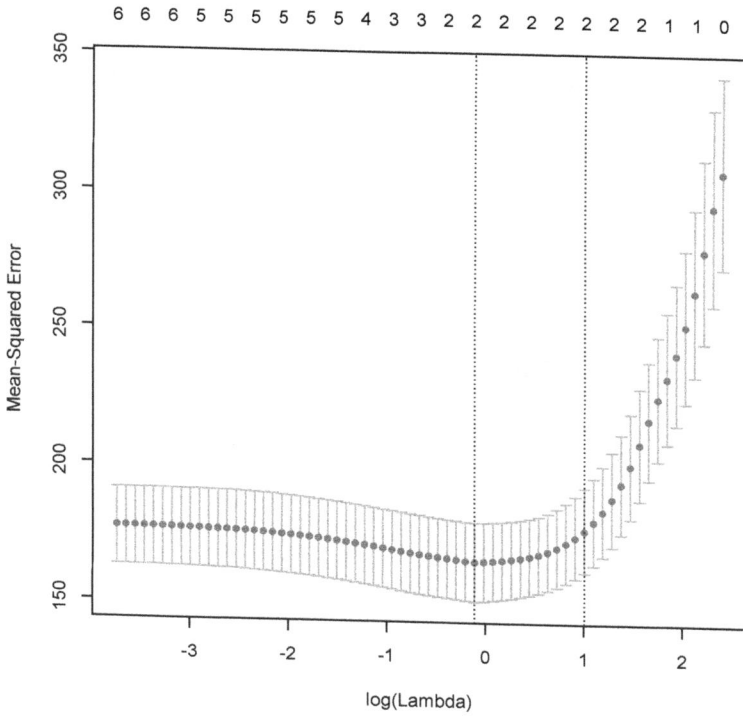

Fig. 11.1   LASSO Regression

Table 11.2 - Regression Parameters under Ridge

Constant	Cement	BFS	FA	Water	SP	CA
-14.915	-0.013	-0.097	-0.009	0.466	-0.071	-0.016

Finally we look at the splines with Compressive strength as the output variable and Cement as the input variable. We use the 'lm' function of the splines package in R to fit the cubic spline function.

Figure 11.3 gives the scatter diagram for the two variables and cubic splines spline with knots as 180, 240 and 300.

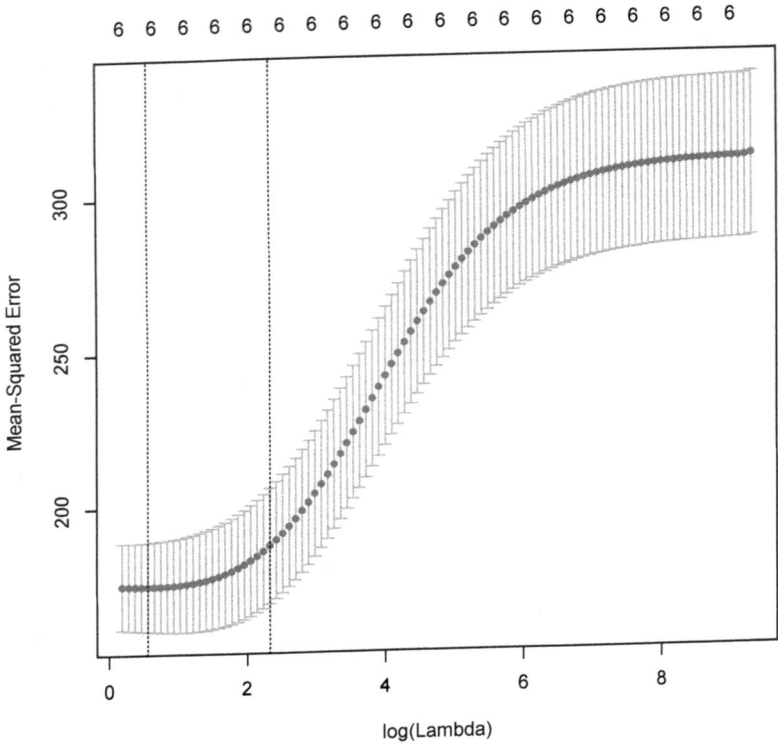

Fig. 11.2   Ridge Regression

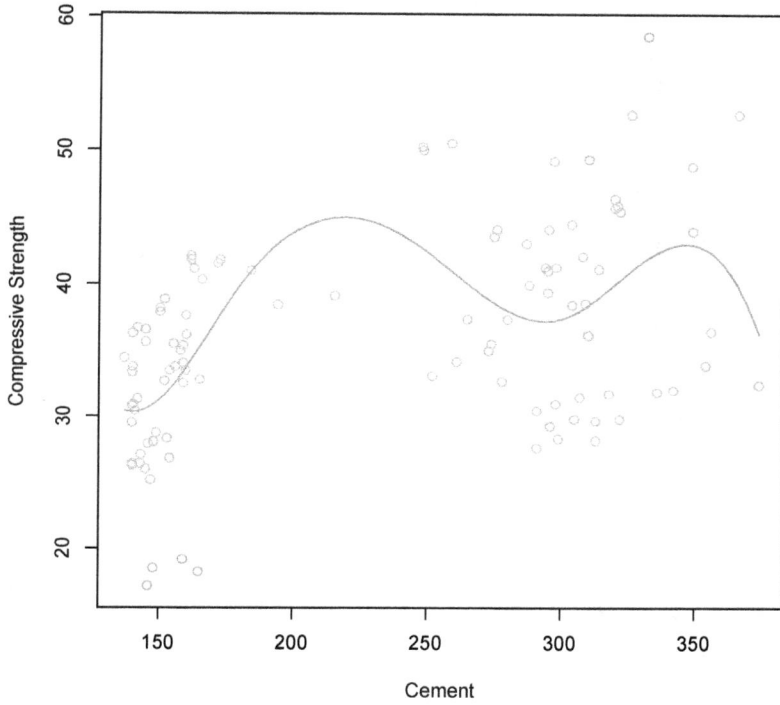

Fig. 11.3   Spline

Chapter 12

NONPARAMETRIC BAYESIAN METHODS

We introduced Bayesian methods for the parametric setting in Section 1.9, which incorporate the preferences of the statistician/scientist about the parameter through a probability distribution on the parameter space. The parameter space in the nonparametric setting is a set of functions such as the set of all distribution functions or the set of all density functions and is thus infinite dimensional. The Bayesian approach, therefore, involves specifying priors, that is define probability measures, on an infinite dimensional space.

12.1 Estimation of the Distribution Function

Suppose we want to obtain an estimator of the c.d.f. F. Let P denote the corresponding probability measure. One may have some prior belief about the c.d.f. based on past data or expert opinion. For example, one can have some idea about the distribution of marks for a certain course or the distribution of the life lengths of certain systems. A c.d.f. attaches probabilities to sub-intervals of R. Suppose from the past data we have some estimates of the probabilities of a fairly large number of sub-intervals of R. These probabilities could be the average probabilities of the intervals estimated from different data sets from the same population. We thus also have an estimate of the variance of these probabilities. Let $P_0(A)$ denote the average probability of an interval A. Then our prior estimate of $P(A)$ is $P_0(A)$. Suppose now we obtain a random sample $X_1, \cdots, X_n$ from P and wish to update our prior estimator. Without prior knowledge, the e.d.f. F_n, defined in Chapter 3 is taken as a reasonable estimator of F. Having a prior estimator and a data based estimator, a simple estimator combining both is an average of the two. The current estimator is based on a sample of size n, whereas the prior could be based on different sized data sets. Let α correspond to the average sample size on which the prior estimate is based. We may take the new estimator to be:

$$\hat{F}_n(t) = \frac{\alpha}{\alpha + n} P_0((-\infty, t]) + \frac{n}{n + \alpha} F_n(t). \qquad (12.1.1)$$

The above estimator is precisely the Bayes estimator obtained by using the Dirichlet prior introduced by [Ferguson (1973)] on the space of probability measures on R. We define this prior below. First let us consider some measure theoretic preliminaries.

12.1.1   *Measure Theoretic Preliminaries*

Let $\mathbf{M}(R)$ denote the space of all probability measures on $\mathbf{B}$, the Borel σ-algebra of subsets of R. To define a probability measure on $\mathbf{M}(R)$ we need to consider a σ-algebra $\mathcal{B}$ of subsets of $\mathbf{M}(R)$. The σ-algebra $\mathcal{B}$ on $\mathbf{M}(R)$ is taken to be the smallest $\sigma-$algebra with respect to which the functions $\{P \to P(B) | B \in \mathbf{B}\}$ from $\mathbf{M}(R)$ to R are measurable.

Let $(\Omega, \mathcal{F}, \mathcal{P})$ be a probability space. A function P defined on $(\Omega, \mathcal{F}, \mathcal{P})$ and taking values in $\mathbf{M}(R)$ is called a random probability measure . That is, for each $\omega \in \Omega$, $P = P(\omega, \cdot)$ is a probability measure on $(R, \mathbf{B})$. We note that the σ-algebra $\mathcal{B}$ has been chosen so that for a random probability measure $P(w, .)$, $P(., B)$ is a random variable from $(\Omega, \mathcal{F})$ to $(R, \mathbf{B})$ for each Borel subset B of R. A prior then can be considered as the probability law of a random probability measure or a stochastic process indexed by the sets in the Borel $\sigma-$algebra $\mathbf{B}$. Therefore to define a prior we need to specify a probability distribution $\Pi_{B_1, \cdots, B_k}$ of $(P(B_1), \cdots, P(B_k))$ for each finite partition $(B_1, \cdots, B_k)$ of R, by Borel sets so that the Kolmogorov consistency ([Billingsley (1995)], p. 486) conditions hold. That is,

(i) $\Pi_{B_1, \cdots, B_k}$ is a probability measure on $S_k = \{(x_1, \cdots, x_k), 0 \leq x_i \leq 1, \sum_{i=1}^{i=k} x_i \leq 1\}$, the k dimensional simplex.

(ii) If $(A_1, \cdots, A_m)$ is another collection of disjoint Borel subsets, whose elements are unions of sets from $(B_1, \cdots, B_k)$, then $\Pi_{A_1, \cdots, A_m}$ is the distribution of

$$\left(\sum_{B_i \subset A_1} P(B_i), \cdots, \sum_{B_i \subset A_m} P(B_i) \right),$$

(iii) if $B_n \downarrow \emptyset$ then $\prod_{B_n}$ converges weakly to a distribution degenerate at 'zero', and

(iv) $P(R) = 1$, $a.s.$.

For a rigorous treatment, we refer to ([Ghosh and Ramamoorthi (2003)], Chapter 2).

Next we define what is meant by a random sample from P for a random probability measure P [Ferguson (1973)].

Definition 12.1.1. Suppose P is a random probability measure on $(R, \mathbf{B})$. We say that conditional on P, the random variables $X_1, \cdots, X_n$, defined on the probability space $(\Omega, \mathcal{F}, \mathcal{P})$, form a random sample of size n from P if for each integer $m \geq 1$, and for all Borel sets $A_1, \cdots, A_m, B_1, \cdots, B_n$,

$$\mathcal{P}(X_1 \in B_1, \cdots, X_n \in B_n | P(A_1), \cdots, P(A_m), P(B_1), \cdots, P(B_n))$$

$$= \prod_{i=1}^{n} P(B_i), \quad a.s. \ \mathcal{P}.$$

12.2 The Dirichlet Prior

We first define the Dirichlet distribution.

Definition 12.2.1. The random vector $(X_1, \cdots, X_{k-1}, X_k)$, where $X_k = 1 - \sum_{i=1}^{k-1} X_i$, is said to have the Dirichlet distribution of order $k \geq 2$ with parameters $(\alpha_1, \cdots, \alpha_k)$ if the density (with respect to the $k-1$ dimensional Lebesgue measure) of the distribution of $(X_1, \cdots, X_{k-1})$ is

$$f(x_1, \cdots, x_{k-1}) = \frac{\Gamma(\sum_{i=1}^{k} \alpha_i)}{\prod_{i=1}^{k} \Gamma(\alpha_i)} (\prod_{i=1}^{k-1} x_i^{\alpha_i - 1})(1 - \sum_{i=1}^{k-1} x_i)^{\alpha_k - 1},$$

on the $k-1$ dimensional simplex $S_{(k-1)}$ and is *zero* elsewhere and where $\alpha_i > 0$, $x_i > 0$, $i = 1, \cdots, k$. We denote the Dirichlet distribution of order k with parameter $(\alpha_1, \cdots, \alpha_k)$ by $D(\alpha_1, \cdots, \alpha_k)$.

Note that the distribution of $(X_1, \cdots, X_{k-1}, X_k)$ is singular with respect to the Lebesgue measure on the k-dimensional space since $X_k = 1 - \sum_{i=1}^{i=k-1} X_i$. For $k = 2$, $D(\alpha_1, \alpha_2)$ corresponds to the Beta distribution with support $(0, 1)$ and parameters (α_1, α_2). The Dirichlet distribution with parameters $(\alpha_1, \cdots, \alpha_k)$, with $\alpha_i \geq 0$, and $\alpha_j = 0$ for $j \in J \subset \{1, 2, \cdots, k\}$, is the distribution of the vector $(X_1, \cdots, X_{k-1}, X_k)$ such the $X_j = 0$ for $j \in J$ and such that $(X_i; i \notin J)$ has a lower dimensional Dirichlet distribution with density as given in the above definition.

Two properties of interest of the Dirichlet distribution are:

(1) If the distribution of $(X_1, \cdots, X_k)$ is $D(\alpha_1, \cdots, \alpha_k)$ and $s = \sum_{i=1}^{k} \alpha_i$ then the marginal distribution of X_j is $Beta(\alpha_j, s - \alpha_j)$. Thus $E(X_i) = \alpha_i/s$, $Var(X_i) = \alpha_i(\alpha - \alpha_i)/(s^2(s+1))$, and $cov(X_i, X_j) = \alpha_i \alpha_j/(s^2(s+1))$, $i \neq j$.

(2) Let the distribution of $(X_1, \cdots, X_k)$ be $D(\alpha_1, \cdots, \alpha_k)$. If $n_1, \cdots, n_m$ are integers such that $0 < n_1 < \cdots < n_m = k$ then the distribution of $(\sum_{i=1}^{n_1} X_i, \sum_{i=n_1+1}^{n_2} X_i, \cdots, \sum_{i=n_{m-1}+1}^{n_m} X_i)$ is $D(\sum_{i=1}^{n_1} \alpha_i, \sum_{i=n_1+1}^{n_2} \alpha_i, \cdots, \sum_{i=n_{m-1}+1}^{n_m} \alpha_i)$.

Definition 12.2.2. Let P_0 be a probability measure on $(R, \mathbf{B})$ and α a non-negative real number. A random probability measure P on $(R, \mathbf{B})$ is said to be distributed as a Dirichlet process with parameter (α, P_0) if for every finite partition $B_1, \cdots, B_m$ of R by Borel sets, the joint distribution of $(P(B_1), \cdots, P(B_m))$ is Dirichlet $D(\alpha P_0(B_1), \cdots, \alpha P_0(B_m))$.

The existence of the Dirichlet Process follows from the Theorem 3.2.1 ([Ghosh and Ramamoorthi (2003)], p. 96). We will denote a Dirichlet process with parameter (α, P_0) by $\mathcal{D}_{\alpha, P_0}$.

Comments:
(i) For a Borel subset B of R, by considering the partition B, B^c, we see that the $P(B)$ has the $Beta(\alpha P_0(B), \alpha - \alpha P_0(B))$ distribution. Therefore, the expectation of $P(B)$, that is, $\mathcal{E}(P(B)) = P_0(B)$. Thus the parameter P_0, called the base measure, does have the connotation as the average probability. It is the mean of the process $\mathcal{D}_{\alpha, P_0}$.
(ii) Also note that $Var(P(B)) = P_0(B)(1 - P_0(B))/(\alpha + 1)$, which shows that the variance of the prior distribution decreases as α increases, therefore α is referred to as the concentration parameter. Since as $\alpha \to \infty$, $P(B)$ approaches, in probability, to its mean $P_0(B)$, α may also be considered to represent the 'sample size' of the prior with reference to 'the law of large numbers'.
(iii) Let P be a random probability measure on $(R, \mathbf{B})$ and $F(t) = P((-\infty, t])$ the corresponding random c.d.f. Then F is a Dirichlet process with parameter (α, F_0) iff P is a Dirichlet process with parameter (α, P_0), where $F_0(t) = P_0((-\infty, t])$.

An equivalent definition of the Dirichlet process given by [Sethuraman (1994)] shows that the Dirichlet process gives mass 1 to the set of discrete measures. This definition, also called the stick-breaking algorithm, is helpful for simulating a Dirichlet process. We give the definition below. Let

(iv) $P(R) = 1$, $a.s.$.

For a rigorous treatment, we refer to ([Ghosh and Ramamoorthi (2003)], Chapter 2).

Next we define what is meant by a random sample from P for a random probability measure P [Ferguson (1973)].

Definition 12.1.1. Suppose P is a random probability measure on $(R, \mathbf{B})$. We say that conditional on P, the random variables $X_1, \cdots, X_n$, defined on the probability space $(\Omega, \mathcal{F}, \mathcal{P})$, form a random sample of size n from P if for each integer $m \geq 1$, and for all Borel sets $A_1, \cdots, A_m, B_1, \cdots, B_n$,

$$\mathcal{P}(X_1 \in B_1, \cdots, X_n \in B_n | P(A_1), \cdots, P(A_m), P(B_1), \cdots, P(B_n))$$
$$= \prod_{i=1}^{n} P(B_i), \quad a.s. \ \mathcal{P}.$$

12.2 The Dirichlet Prior

We first define the Dirichlet distribution.

Definition 12.2.1. The random vector $(X_1, \cdots, X_{k-1}, X_k)$, where $X_k = 1 - \sum_{i=1}^{k-1} X_i$, is said to have the Dirichlet distribution of order $k \geq 2$ with parameters $(\alpha_1, \cdots, \alpha_k)$ if the density (with respect to the $k-1$ dimensional Lebesgue measure) of the distribution of $(X_1, \cdots, X_{k-1})$ is

$$f(x_1, \cdots, x_{k-1}) = \frac{\Gamma(\sum_{i=1}^{k} \alpha_i)}{\prod_{i=1}^{k} \Gamma(\alpha_i)} (\prod_{i=1}^{k-1} x_i^{\alpha_i - 1})(1 - \sum_{i=1}^{k-1} x_i)^{\alpha_k - 1},$$

on the $k - 1$ dimensional simplex $S_{(k-1)}$ and is *zero* elsewhere and where $\alpha_i > 0$, $x_i > 0$, $i = 1, \cdots, k$. We denote the Dirichlet distribution of order k with parameter $(\alpha_1, \cdots, \alpha_k)$ by $D(\alpha_1, \cdots, \alpha_k)$.

Note that the distribution of $(X_1, \cdots, X_{k-1}, X_k)$ is singular with respect to the Lebesgue measure on the k-dimensional space since $X_k = 1 - \sum_{i=1}^{i=k-1} X_i$. For $k = 2$, $D(\alpha_1, \alpha_2)$ corresponds to the Beta distribution with support $(0, 1)$ and parameters (α_1, α_2). The Dirichlet distribution with parameters $(\alpha_1, \cdots, \alpha_k)$, with $\alpha_i \geq 0$, and $\alpha_j = 0$ for $j \in J \subset \{1, 2, \cdots, k\}$, is the distribution of the vector $(X_1, \cdots, X_{k-1}, X_k)$ such the $X_j = 0$ for $j \in J$ and such that $(X_i; i \notin J)$ has a lower dimensional Dirichlet distribution with density as given in the above definition.

Two properties of interest of the Dirichlet distribution are:

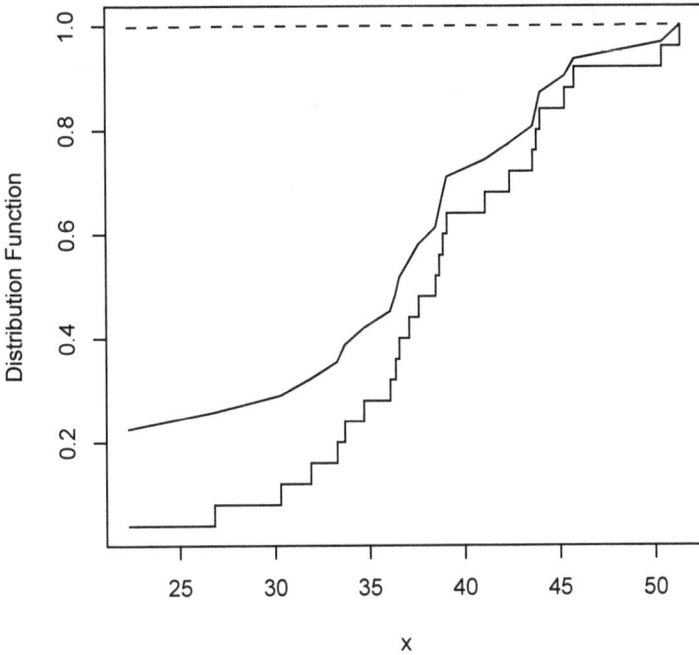

Fig. 12.3   Empirical and Bayes cdf

that is,

$$P_0((-\infty, m_0)) \leq 1/2 \leq P_0((-\infty, m_0]). \qquad (12.2.5)$$

Now since m is the median of F, $\mathcal{P}[m < m_0] = \mathcal{P}[F(m_0) > 1/2]$ and $\mathcal{P}[m \leq m_0] \leq \mathcal{P}[F(m_0) \geq 1/2]$. But the distribution of $F(m_0)$ is $Beta(\alpha P_0((-\infty, m_0]), \alpha P_0((m_0, \infty)))$. Thus (12.2.4) holds if and only if the median of this Beta distribution is $1/2$, which holds if and only if the two parameters of the Beta distribution are equal and which holds if and only if (12.2.5) holds. This establishes that the median of the distribution of m, is the same as the median of P_0.

As mentioned earlier, to obtain the posterior estimator given a sample of size n, we replace (α, P_0) by $(\alpha + n, \frac{\alpha P_0 + \sum_{i=1}^{n} \delta_{x_i}}{\alpha + n})$. Thus the Bayes estimator with respect to the absolute error loss function is the median of $\hat{F}_n$

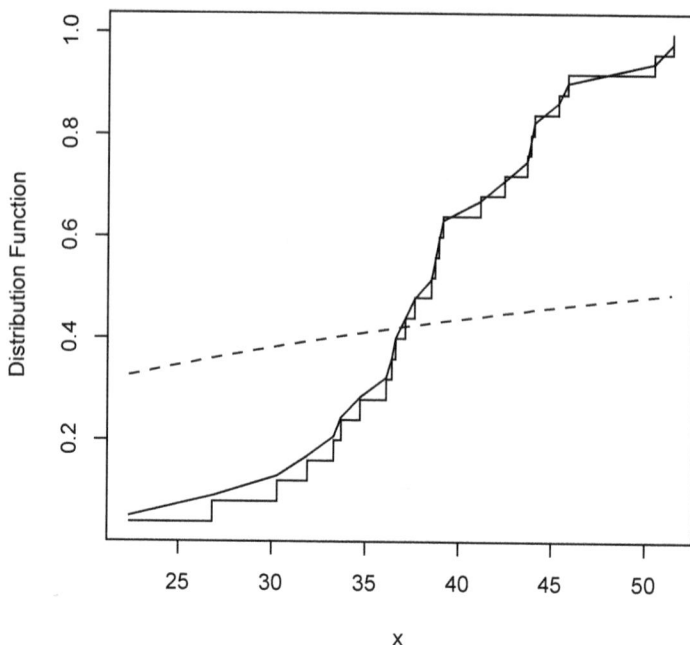

Fig. 12.2 Empirical and Bayes cdf

2. *Estimate of the median* ([Ferguson (1973)]): Suppose the problem is to estimate the median m of an unknown probability measure P on $(R, \mathbf{B})$. We consider the Dirichlet prior with parameter (α, P_0) on $\mathbf{M}(R)$. The median m of a random probability distribution F is random. The Bayes estimate of m under the squared error loss and the no-sample problem is the expectation of m, which is difficult to compute. For the no-sample problem, the Bayes estimate of m under the absolute error loss function is the median of the distribution of m, which is the same as the median of $\mathcal{E}(P) = P_0$. To see this, we first note that a number m_0 is a median of the distribution of m if and only if

$$\mathcal{P}[m < m_0] \leq 1/2 \leq \mathcal{P}[m \leq m_0]. \qquad (12.2.4)$$

To show that m_0 is a median of $\mathcal{E}(P)$, we need to show that

$$\mathcal{E}(P(-\infty, m_0)) \leq 1/2 \leq \mathcal{E}(P(-\infty, m_0]),$$

$$\Pi(t|\theta_{-i}, X_i) = \frac{\int_{-\infty}^{t} k(X_i, \theta) d(\alpha G_0 + \sum_{j=1, j\neq i}^{n} \delta_{\theta_j})(\theta)}{\int k(X_i, \theta) d(\alpha G_0 + \sum_{j=1, j\neq i}^{n} \delta_{\theta_j})(\theta)}.$$

Let $q_{0,i} = \int k(X_i, \theta) dG_0(\theta)$, and $q_{i,j} = k(X_i, \theta_j)$. Then the denominator of the above expression is $\alpha\, q_{0,i} + \sum_{j=1, j\neq i}^{n} q_{i,j}$. Let g_0 denote the density corresponding to the base c.d.f. G_0 and $h(\theta_i|X_i)$ denote the density of θ_i given X_i. Then $h(\theta_i|X_i) = k(X_i, \theta_i) g_0(\theta_i)/q_{0,i}$, and

$$\Pi(t|\theta_{-i}, X_i) = \frac{\alpha\, q_{0,i} \int_{-\infty}^{t} h(\theta|X_i) d\theta + \sum_{j=1, j\neq i}^{n} q_{i,j} I_{[\theta_j, \infty)}(t)}{\alpha\, q_{0,i} + \sum_{j=1, j\neq i}^{n} q_{i,j}}. \qquad (12.3.3)$$

Algorithm for the Gibbs sampler

Step 1: Choose initial values of the elements of $\underline{\theta}$, these could be samples from the posterior distribution $H(\theta_i|X_i)$ or could be generated from the marginal distribution G_0. Let us denote these values by $\underline{\theta}^{(0)}$.

Step 2: Sample elements of $\underline{\theta}$ sequentially by drawing $\theta_1^{(1)}$ from the distribution of $(\theta_1|\theta_{-1}^{(0)}, X_1)$, then $\theta_2^{(1)}$ from the distribution of $(\theta_2|\theta_1^{(1)}, \theta_j^{(0)}, \{j = 3, \cdots, n\}, X_2)$, $\theta_3^{(1)}$ from the distribution of $(\theta_3|\theta_1^{(1)}, \theta_2^{(1)}, \theta_j^{(0)}, \{j = 4, \cdots, n\}, X_3)$ and so on up to $\theta_n^{(1)}$ from the distribution of $(\theta_n|\theta_j^{(1)}, \{j = 1, \cdots, n-1\}, X_n)$.

Step 3: Return to Step 2 and proceed iteratively until convergence.

The above sampling process results in approximate samples from the posterior of $\underline{\theta}$ given the data. Since G is discrete, multiple θ_i's can have the same value, which induces a clustering of the θ_i's. Thus the mixture will reduce to less than n components. Let n^* denote the number of distinct elements of the vector $\underline{\theta}$ and θ_j^*, $j = 1, \cdots, n^*$, the distinct θ_i's. We can say that θ_j^* represents the cluster j. Let $\underline{w} = (w_1, \cdots, w_n)$, where the w_i's indicate the cluster, that is, $w_i = j$ if and only if $\theta_i = \theta_j^*$, $i = 1, \cdots, n$. Then for each $i = 1, \cdots, n$ the conditional distribution of θ_i given θ_{-i} and the data is

$$\frac{\alpha\, q_{0,i} \int_{-\infty}^{t} h(\theta|X_i) d\theta + \sum_{j=1, j\neq i}^{n_i^{*-}} n_{i,j}^- q_{i,j}^* I_{[\theta_j^*, \infty)}(t)}{\alpha\, q_{0,i} + \sum_{j=1, j\neq i}^{n_i^{*-}} q_{i,j}^*},$$

where n_i^{*-} denotes the number of clusters in θ_{-i}, $n_{i,j}^-$ the number of elements in the j-th cluster of θ_{-i} and $q_{i,j}^* = k(X_i, \theta_j^*)$.

Suppose the kernel k involves a hyper-parameter ϕ, the concentration parameter α is random and the base measure G_0 has a hyper-parameter ψ. We, then, have to assume priors for ϕ, α, and ψ. The joint posterior is

$$\Pi(G, \underline{\theta}, \phi, \alpha, \psi|\underline{x}_n) = \Pi(G|\underline{\theta}, \alpha, \psi)\Pi(\underline{\theta}, \phi, \alpha, \psi|\underline{x}_n).$$

The posterior computations for the MDP models with normal kernels and their extensions are implemented in the R package DPpackage (Jara et al. 2011) using the Gibbs sampling approach.

12.3.3   *Mixture of Normal Kernels*

In the above formulation of the MDP, it is assumed that $k(\cdot, \theta_i)$ is the density of a normal r.v. with mean μ_i and variance V_i; with $\theta_i = (\mu_i, V_i)$. The θ_i come from a prior distribution G on $\Theta = R \times R^+$. The G is modeled as a $Dirichlet(\alpha, G_0)$ process where α is a positive scalar and $G_0(\cdot)$ is a specified bivariate distribution function over Θ.

The function DPdensity in the R package DPpackage generates a posterior density sample for a MDP normal model.

The baseline distribution G_0 is taken to be the normal-inverted-Wishart distribution, which is conjugate prior of a multivariate normal distribution with unknown mean and covariance matrix. That is,

$$G_0 = N(\mu|m_1, (1/k_0)V)IW(V|\nu_1, \psi_1).$$

The package has the choice of considering independent hyperpriors. The concentration parameter α given a_0, b_0 has the $Gamma(a_0, b_0)$ distribution, m_1 given m_2, s_2 is $N(m_2, s_2)$, k_0 given τ_1, τ_2 is $Gamma(\tau_1/2, \tau_2/2)$ and ψ_1 given ν_2, ψ_2 is $IW(\nu_2, \psi_2)$. The inverted-Wishart prior is parametrized such that if A follows $IW_q(\nu, \psi)$ then $E(A) = \psi^{-1}/(\nu - q - 1)$. If the parameters of G_0 are fixed, the corresponding hyperparameters should be set to NULL in the hyper-parameter specification of the model.

Choice of the parameters of the prior
The following discussion is based on [Ferguson (1983)] and [Escobar and West (1995)] and is just an aid for choosing the prior parameters. We consider the kernel $k(\cdot, \theta_i)$ to be the density of a normal r.v. with mean μ_i and variance V_i; with $\theta_i = (\mu_i, V_i)$. For the base distribution G_0, we assume that the conditional distribution of μ_i given V_i is $N(m_1, V_i/k_0)$ for some mean m_1 and scale fator $1/k_0$, and the distribution of V_i is one-dimensional inverse-Wishart(ν_1, ψ_1), that is inverse-gamma$(\nu_1/2, 1/(2\psi_1))$. We note that the inverse-Wishart distribution mentioned in the DPpackage is a multivarite generalization of the inverse-gamma distribution. We either have to choose values for the parameters m_1, k_0, ν_1, ψ_1 and the concentration parameter α or choose priors for these. Since $f_0(x)$, the

'no sample' guess of $f(x)$ is the expectation of $f(x)$ under the prior distribution, we have

$$f_0(x) = \mathcal{E}(f(x)) = \int_{\Theta} k(x,\theta)dG_0(\theta)$$

$$= \frac{\Gamma(\nu_1/2 + 1/2)}{\Gamma(\nu_1/2)\Gamma(1/2)}\sqrt{\frac{\psi_1}{(1+k_0)}}\left(1 + \frac{\psi_1}{(1+k_0)}(x - m_1)^2\right)^{-(\nu_1/2+1/2)}.$$

The mean of $f_0(.)$ is m_1 and the variance is $\frac{(1+k_0)}{\psi_1(\nu_1+2)}$. Thus choose m_1 to be either the value of the belief about the centre of mass or the sample mean. The mean of μ_i is m_1 and $Var(\mu_i) = E(V_i)/k_0$. If the uncertainty in the value of μ_i is greater than (equal to, less than) the average variance, then k_0 should be chosen less than (equal to, greater than) 1. Now $E(V_i) = 1/(\psi_1(\nu_1-2))$. Thus the values for the parameters may be chosen to match the choice of k_0 and the prior belief about the variance of the observations.

From (12.3.2) we see that, the influence of the data comes in only through the second term and the first term contains only the influence of the prior. Therefore a small value for α should be chosen, if one is not too sure about the prior. Further, from (12.3.3), we see that θ_i will be a new distinct value with probability $\alpha q_{0i}/(\alpha q_{0i} + \sum_{j\neq i} q_{ij})$ and is otherwise one of the existing $(n-1)$ values. Therefore, a small value of α causes larger number of θ_i's to be identical, which will result in a small number of mixture components or clusters.

For given n^* distinct values among the elements of $\underline{\theta}$, a small value of k_0 implies a larger dispersion among the n^* group means μ_j^*, which for a fixed V_j^* leads to a greater chance of multimodality in the predictive distribution. That is, we should choose a small value for k_0, if we feel that the density to be estimated is multimodal. The value of k_0 is related to the smoothing like the bin-width is in the kernel density estimation discussed in Chapter 10.

Example 12.3: Figure 12.5 shows the plot of the (posterior) density estimator for the data given in the Example 12.2. We assume a normal MDP prior and use the DPdensity function of the DPpackage. We chose the concentration/precision parameter $\alpha = 1$, the mean $m_1 = 38$, the mean of the data. The average value of the parameter k_0 was taken to be 5. We chose $\tau_1 = 500$ and $\tau_2 = 100$, so that k_0 has a small variance. Since the variance of f_0 is $(k_0+1)/(\psi_1(\nu_1+2))$, we equated this expression to 46, the variance of the data. We took $\nu_1 = 4$ and thus $1/\psi_1 = 46$. The histogram of the data and the plot of the kernel density estimator (dashed line) with the

'Epanechnikov' kernel (see Chapter 10) and bandwidth=1 are also shown on the same graph.

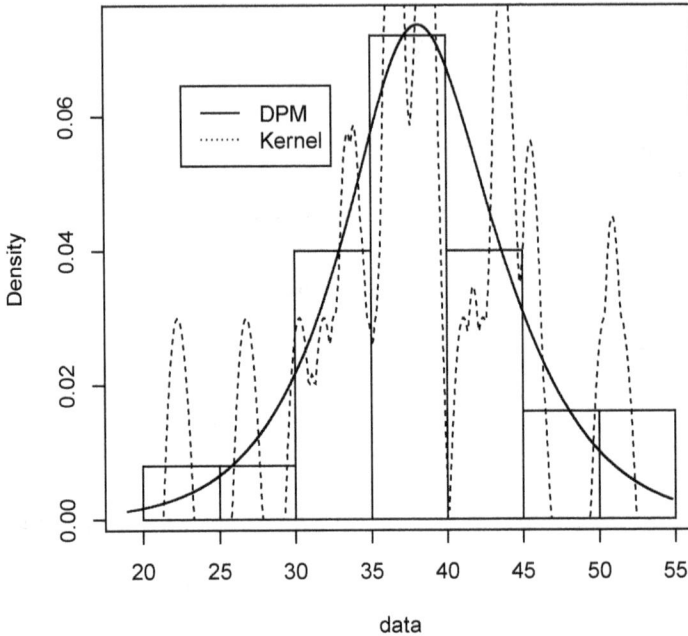

Fig. 12.5   Density estimates

Example 12.4: We generated 25 observations from the Gamma(2,2) density. Figure 12.6 shows the plots of: the true density, its estimate based on the MDP prior, its kernel density estimate and the histogram of the data. Figure 12.7 shows the plots for sample size 100. The parameters of the priors were chosen as in the Example 12.3.

In the above package, a function called DPcdensity is also included, which generates a posterior density sample for a Bayesian density regression model with continuous predictors(co-variates) using a MDP normal model.

For lifetime data analysis, [Kottas (2006)] gives a computational approach to obtain the posterior distribution of a MDP using a Weibull kernel. [Cheng and Yuan (2013)] develop an algorithm that implements the Gibbs

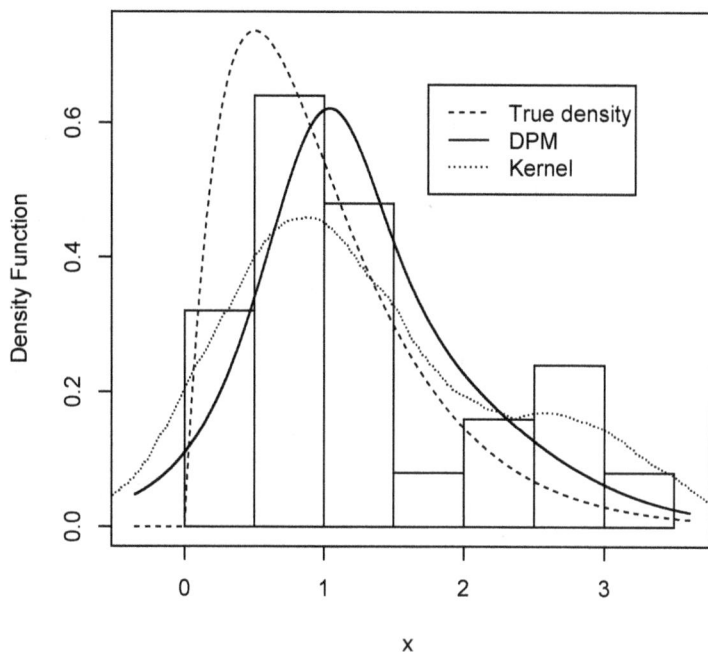

Fig. 12.6   Density estimates

sampling to fit the MDP with a log-normal kernel for a rightly censored failure time data. The choice the kernel $k(x, \theta)$ depends on the underlying sample space. If the underlying density function is defined on the entire real line, a location-scale kernel is appropriate. If the density is known to have support on the unit interval, $k(x, \theta)$ may be taken to be a *Beta* density. For lifetime data analysis, $k(x, \theta)$ may be taken to be gamma, Weibull or lognormal densities. [Petrone and Veronese (2002)] have discussed a way of viewing the choice of a kernel through the notion of a Feller sampling scheme.

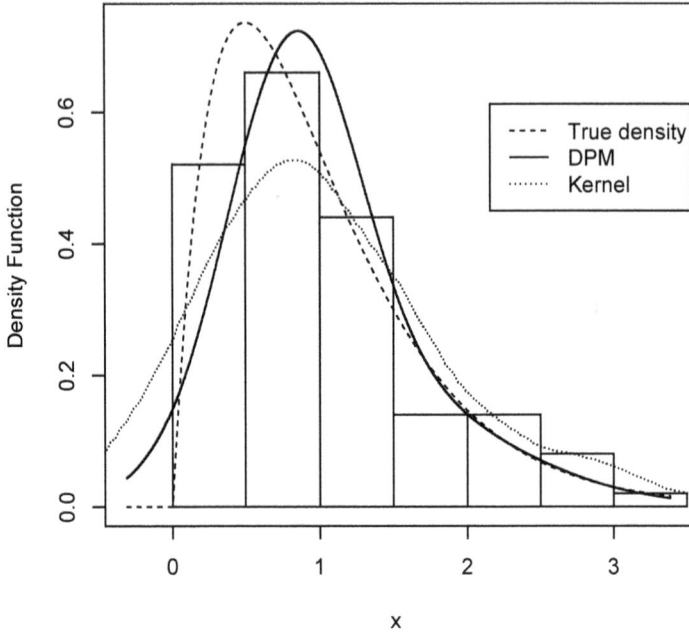

Fig. 12.7   Density estimates

12.4   Other Priors

Some other priors considered are the Pólya Tree Priors and the Neutral to the Right priors. The Pólya tree priors were formulated by [Ferguson (1974)] and later discussed by [Mauldin *et al.* (1992)] and [Lavine (1992)], [Lavine (1994)].

Pólya tree priors are specified through a nested tree of measurable partitions of R. Let $\{\Gamma_k; k \geq 1\}$ be such a nested tree. That is, Γ_{k+1} is a refinement Γ_k (each set in Γ_{k+1} is a union of sets in Γ_k.) Further $\bigcup_{k \geq 1} \Gamma_k$ generates the Borel $\sigma-$algebra. We start with $\Gamma_1 = \{B_0, B_1\}$, $\Gamma_2 = \{B_{00}, B_{01}, B_{10}, B_{11}\}$, where $B_0 = B_{00} \cup B_{01}$, $B_1 = B_{10} \cup B_{11}$. Thus $\Gamma_m = \{B_\varepsilon, \varepsilon = \varepsilon_1 \cdots \varepsilon_m\}$, where $\varepsilon \in \{0,1\}^m$. For $\varepsilon \in \{0,1\}^m$, the sets $B_{\varepsilon 0}$ and $B_{\varepsilon 1}$ are in Γ_{m+1} and $B_\varepsilon = B_{\varepsilon 0} \cup B_{\varepsilon 1}$. Let $E^* = \bigcup_{m \geq 1} \{0,1\}^m$ be the set of all sequences of zeros and ones and let $\mathcal{A} = \{\alpha_\varepsilon, \ \varepsilon \in E^*\}$ be a set of non-negative real numbers.

Definition 12.4.1 (Lavine 1992): A random probability measure P is a Pólya tree process with respect to $\Gamma = \{\Gamma_m, m \geq 1\}$ and $\mathcal{A}$ (denoted by $PT(\mathcal{A}, \Gamma)$), if there exist random variables $\mathcal{Y} = \{Y_\varepsilon, \ \varepsilon \in E^*\}$ such that
(i) All random variables in $\mathcal{Y}$ are mutually independent.
(ii) For every $\varepsilon \in E^*$, Y_ε has a $Beta(\alpha_{\varepsilon 0}, \alpha_{\varepsilon 1})$ distribution.
(iii) For every $\varepsilon = \varepsilon_1 \cdots \varepsilon_m \in \{0, 1\}^m$,
$P(B_{\varepsilon_1 \cdots \varepsilon_m}) = \prod_{j=1}^m (Y_{\varepsilon_1 \cdots \varepsilon_{j-1}})^{1-\varepsilon_j} (1 - Y_{\varepsilon_1 \cdots \varepsilon_{j-1}})^{\varepsilon_j}$, where $\varepsilon_0 = Y$ has the $Beta(\alpha_0, \alpha_1)$ distribution.

With a suitable choice of the parameters, a Pólya Tree prior can be ensured to sit on the space of densities. Both, the Dirichlet prior and the Pólya tree priors are what are known as tail-free processes. A random probability measure P is a tail-free process with respect to a nested partition $\{\Gamma_k; k \geq 1\}$, if the sets of random variables $\{P(B|A), A \in \Gamma_k, B \in \Gamma_{k+1}\}$ for $k = 1, 2, \cdots$ are independent.

Neutral to the right priors are another generalization of Dirichlet Process.

Definition 12.4.2. A random probability measure $P \in M(R)$ or the corresponding random distribution F is said to be neutral to the right (NTR) if for each $k > 1$ and all $t_1 < \cdots < t_k$, there exist independent $[0, 1]$ valued random variables $V_1, V_2, \ldots, V_k$ such that $(F(t_1); \ldots; F(t_k))$ has the same distribution as $(V_1; 1 - (1-V_1)(1-V_2); \ldots; 1 - \prod_1^k (1-V_i))$. The corresponding prior is said to be NTR. In other words, $V_1 = F(1), V_i = \frac{1-F(t_i)}{1-F(t_{i-1})} = Pr(X > t_i | X > t_{i-1})$ are independent.

Neutral to the right priors were introduced by [Doksum (1974)] and further developed by [Hjort (1990)]. They are conjugate in the presence of right censored data, i.e. if we start with a NTR prior then the posterior distribution given data, some of which are right censored is again NTR. These priors too are supported by the set of discrete distributions. The mathematical tractability of NTR priors comes from their intimate relation with independent increment processes. Doksum, after introducing NTR priors, showed a connection between NTR priors and independent increment processes. He considered the cumulative hazard function defined by $\mathbf{D}(F)(t) = -\log(1 - F(t))$, and showed that, a random distribution function F is neutral to the right if and only if $\{\mathbf{D}(F)(t) : t \geq 0\}$ is an independent increment process. Hjort in ([Hjort (1990)]), considered the following formula for the cumulative hazard function, which is defined even when F has no density. He considered, with $T_F = \inf\{t|F(t) = 1\}$,

$$\mathbf{H}(F)(t) = H_F(t) = \begin{cases} \int_{(0,t]} \frac{dF(s)}{F[s,\infty)} & \text{for } t \leq T_F \\ H_F(T_F) & \text{for } t > T_F \end{cases}$$

and showed that in this case too, a prior is NTR iff the cumulative hazard function is distributed as an independent increment process.

As mentioned earlier, the mathematical tractability of NTR priors arise from their connection with independent increment process. Independent increment processes are determined by their so called Levy measure. Starting with a Levy measure, Hjort gives explicit expression for the Levy measure of the posterior. In addition in [Hjort (1990)], he considers a class of priors called Beta Process which correspond to a certain form of Levy measures. The finite dimensional distribution of these process arise as a limit of Beta distributions. For further details we refer to [Hjort (1990)], [Ghosh and Ramamoorthi (2003)], [Dey *et al.* (2003)].

12.5 Regression and Classification

Let Y denote the response (output) variable and $\underline{X}$ the vector of input variables (covariates, explanatory variables, predictors). The estimation of the regression function $g(x) = E[Y|X = x]$ based on the data (y_i, x_i), $i = 1, \cdots, n$ is considered in Chapter 11. We first consider the approach of [Müller *et al.* (1996)], which is the multivariate analogue of the normal mixtures described above. The joint distribution of $(Y_i, \underline{X}_i)$ is specified to be $N(\mu_i, \Sigma_i)$, $i = 1, \cdots, n$. The $\theta_i = (\mu_i, \Sigma_i)$ follow a multivariate prior distribution G. The G is modeled as a $Dirichlet(\alpha, P_0)$ process where α is a positive scalar and P_0 is a specified probability measure over Θ.

The regression model is

$$Y_i = f(X_i) + \epsilon_i,$$

where the ϵ_i, $i = 1, \ldots, n$ are independent r.v.s and are independent of the $X_i's$, $E[\epsilon_i] = 0$ and $Var(\epsilon_i) = \sigma^2 < \infty$, $i = 1, \cdots n$.

12.5.1 *Basis Representation*

Consider a basis representation of the regression function f;

$$f(\cdot) = \sum_k \beta_k \phi_k(\cdot),$$

where $\phi_k(\cdot)$, $k = 1, 2, \cdots$ denote the known basis functions, such as polynomials, splines, orthonormal series, wavelets, neural networks, and regression trees.

A prior probability model is defined on the sequences $\{\beta_k,\ k = 1, \cdots\}$. For example assume that β_k has pdf $p_k(\cdot)$, $k = 0, \cdots, M$ or assume a joint distribution of $(\beta_1, \cdots, \beta_M)$ for some fixed M and $\mathcal{P}(\beta_k = 0) = 1$ for $k > M$. The optimal estimator under the squared error loss is $\hat{f}(t) = \sum_{k=1}^{\infty} E(\beta_k|\underline{X})\phi_k(t)$. We observe that this procedure is more like Bayesian parametric estimation, since only finite dimensional parameters are estimated using finite dimensional prior distributions.

12.6 Posterior Consistency

Let Π stand for the prior distribution and $\Pi(\cdot|data)$ stand for (a version of) the posterior distribution.

Definition 12.6.1. The posterior distribution is said to be consistent at a given θ_0, or (θ_0, Π) is a consistent pair, if for any neighborhood V of θ_0, $\Pi(\theta \notin V|data) \to 0$ (in probability or a.s.) as the size of the data tends to infinity when θ_0 is the true value of the parameter.

In finite dimensional parametric models, consistency holds in general. The situation is somewhat complex in the nonparametric case. For one, there is a variety of neighborhoods U of P that can be considered. For instance, U can be neighborhoods arising from convergence in distribution or it can be neighborhoods arising from the total variation norm. When the neighborhood under consideration corresponds to convergence in distribution, consistency is usually referred to as weak consistency.

It is known that tail free priors are weakly consistent. A similar result does not hold for NTR priors. [Kim and Lee (2001)] gave an example of a NTR prior which fails to be consistent. Sufficient conditions for consistency of NTR priors is given in [Ghosh and Ramamoorthi (2003)] and in [Dey et al. (2003)]. In more general models, like mixture models, conditions for consistency often involve the prior and other smoothness properties of the model. We refer to [Ghosh and Ramamoorthi (2003)] for an introduction to consistency in the nonparametric case.

12.7 Exercises

(1) Suppose X is a r.v. with the $Beta(\alpha, \beta)$ distribution. Show that the distribution of $(X, 1-X)$ is singular with respect to the two dimensional

Lebesgue measure on R^2.

(2) Consider the problem of estimating the mean μ of an unknown probability measure P on $(R, \mathbf{B})$. Show that with a Dirichlet(α, F_0) prior on $\mathbf{M}(R)$, based on a random sample $X_1, \cdots, X_n$, the Bayes estimator $\hat{\mu}_n$, of μ under the squared error loss function is

$$\hat{\mu}_n = \frac{\alpha}{\alpha + n} \int x dF_0(x) + \frac{n}{\alpha + n} \sum_{i=1}^{n} X_i/n.$$

(3) Generate a sample of size 25 from a known distribution.
(i) Assuming a Dirichlet process prior, obtain a Bayes estimate of the c.d.f. and plot this estimate against the true c.d.f.
(ii) Repeat the simulation 100 times and summarize the results.
(iii) Comment on the performance of the Bayes estimator.

(4) Generate data (sample size 100) from a mixture of known densities.
(i) Estimate the density using the frequentist methods of Chapter 10.
(ii) Estimate the density using the Bayesian methods.
(iii) Plot the density estimates against the true density.
(iv) Compare the estimates.

(5) Suppose conditional on P, $X_1, \cdots, X_n$ is a random sample of size n from P, and P is $\mathcal{D}_{\alpha, P_0}$. Let $(B_1, \cdots, B_m)$ be a partition of R by Borel sets and let $N_k = \#\{i, \ X_i \in B_k\}$. Let $(A_1, \cdots, A_k)$ be another partition such that each A_i is a union of some B_j's, (i.e., $(B_1, \cdots, B_m)$ is a finer partition). Show that the posterior distribution of $(P(A_1), ..., P(A_k))$ given $N_1, \cdots, N_m$ is same as the posterior distribution of $(P(A_1), ..., P(A_k))$ given the counts in $(A_1, \cdots, A_k)$. (Thus we can conclude that the conditional distribution of $(P(B_1), \cdots, P(B_m))$ given the vector $(N_1, \cdots, N_m)$ of cell counts, is same as the conditional distribution of $(P(B_1), \cdots, P(B_m))$ given $(X_1, \cdots, X_n)$.)

APPENDIX

A.1 Introduction

If the data is a sample from a known probability distribution then the sampling distribution of a statistic can be found, at least in principle, through the usual methods of transformation, etc. But when we lack the knowledge of the precise distribution under the nonparametric models which we have here, the sampling distributions are usually not available. Also some statistics, such as those based on ranks, may have known exact probability distributions for the finite sample sizes. But these become extremely cumbersome to derive as the sample size increases even moderately. In these circumstances we may try to get hold of asymptotic distributions of these statistics as the sample size tends to infinity. These asymptotic distributions then may be used as approximations in place of exact distributions which may not be available or may be tedious to obtain.

Also, many properties like consistency, strong consistency, asymptotic relative efficiency, etc. are defined in the limiting case as the sample size tends to infinity. Hence to discuss these asymptotic properties we need to study the asymptotic distributions of these statistics. In this appendix we shall concentrate on a few limiting theorems and properties which are useful in studying nonparametric inference.

A.2 Laws of Large Numbers

This section states the laws of large numbers that are most often used. The proofs may be found in the books on Probability Theory.

Theorem A.2.1 (Bernoulli's (weak) law of large numbers): *Let n independent trials of an experiment be performed each having probability p*

for an event A. Let $n(A)$ be the number of outcomes resulting in the event A. Then

$$\frac{n(A)}{n} \to p \text{ in probability as } n \to \infty.$$

Theorem A.2.2 (Poisson's (weak) law of large numbers): *Let $X_k,\ k = 1, 2, \dots$ be a sequence of independent random variables with the same mean μ and same variance σ^2. Let $S_n = X_1 + X_2 + \cdots + X_n$. Then*

$$\bar{X}_n = \frac{S_n}{n} \to \mu \text{ in probability as } n \to \infty.$$

Theorem A.2.3 (Borel's (strong) law of large numbers): *Let n independent trials be performed each having probability p $(0 < p < 1)$ for an event A. Let $n(A)$ be the number of outcomes resulting in the event A. Then*

$$\frac{n(A)}{n} \to p \text{ with probability } 1 \text{ as } n \to \infty.$$

Theorem A.2.4 (Kolmogorov's (strong) law of large numbers): *Let $X_k,\ k = 1, 2, \dots$ be a sequence of independent and identically distributed random variables. Then*

$$\bar{X}_n \to \mu \text{ with probability } 1 \text{ as } n \to \infty$$

if and only if μ is the common expectation of X_k.

A.3   Convergence in Distribution

We state below some results from probability theory useful in obtaining asymptotic distributions.

Theorem A.3.1. *Let $\{X_n, n \geq 1\}$ and $\{Y_n, n \geq 1\}$ be two sequences of random variables such that as*

$$X_n - Y_n \xrightarrow{p} 0 \text{ and } X_n \xrightarrow{d} X \text{ as } n \to \infty.$$

Then

$$Y_n \xrightarrow{d} X \text{ as } n \to \infty.$$

Theorem A.3.2 (Slutsky's theorem): *Let $X_1, X_2, \cdots$ be an arbitrary sequence of random variables converging in distribution to a random variable X. Let $Y_1, Y_2, \cdots$ be another sequence of random variables converging in probability to a constant c. Then,*

$$\frac{X_n}{Y_n} \text{ converges in distribution to } \frac{X}{c} \text{ as } n \to \infty \text{ if } c \neq 0.$$

A.4   Central Limit Theorems

This section states the most important Central Limit Theorems (C.L.T.) for sequences of independent random variables. Throughout the section $N(0, 1)$ denotes a standard normal variable and $\Phi(t)$ its c.d.f. The symbols $o(1)$ ($o_p(1)$) mean that the sequence of random variables converges to zero a.s. (in probability) as $n \to \infty$ and the symbols $O(1)$ ($O_p(1)$) mean that the sequence is bounded (bounded in probability).

Theorem A.4.1 (The Lindberg-Lévy theorem): *Let X_k, $k = 1, 2, \ldots$ be a sequence of independent and identically distributed random variables with common mean μ and variance σ^2 ($\sigma^2 < \infty$). Then*

$$Z_n = \sqrt{n} \left(\frac{\bar{X}_n - \mu}{\sigma} \right) \overset{d}{\to} N(0, 1) \text{ as } n \to \infty.$$

Theorem A.4.2 (The Lindeberg Theorem): *Let X_k, $k = 1, 2, \ldots$ be a sequence of independent random variables with means zero, variances $\{\sigma_k^2\}$ and c.d.f.s $\{F_k\}$. Let $S_n = \sum_{i=1}^n X_i$ and $s_n^2 = \sum_{i=1}^n \sigma_i^2$. Suppose*

$$\lim_{n \to \infty} s_n^{-2} \sum_{i=1}^n \int_{[|x| > \varepsilon s_n]} x^2 dF_i = 0, \text{ for all } \varepsilon > 0,$$

(i.e. the Lindeberg condition holds.) Then as $n \to \infty$,

$$\frac{S_n}{s_n} \overset{d}{\to} N(0, 1) \text{ as } n \to \infty.$$

The following C.L.T. involves a moment condition.

Theorem A.4.3 (The Liapounov Theorem): *Let X_k, $k = 1, 2, \ldots$ be a sequence of independent random variables with means zero, variances $\{\sigma_k^2\}$ and c.d.f.s $\{F_k\}$. Let $S_n = \sum_{i=1}^n X_i$ and $s_n^2 = \sum_{i=1}^n \sigma_i^2$. Suppose*

$$\lim_{n \to \infty} \frac{\sum_{i=1}^n E|X_i|^{2+\delta}}{s_n^{2+\delta}} = 0, \text{ for some } \delta > 0,$$

then

$$\frac{S_n}{s_n} \overset{d}{\to} N(0, 1) \text{ as } n \to \infty.$$

Proof and details of the above theorems can be found in books on probability theory, for example [Billingsley (1995)], [Chow and Teicher (1997)].

Theorem A.4.4 (The Hájek and Šidák Theorem): *Let X_k, $k = 1, 2, \ldots$ be a sequence of independent and identically distributed random variables with common mean μ and variance σ^2 ($\sigma^2 < \infty$). Also, let $a_1, a_2, \ldots$*

be a sequence of real numbers such that

$$\lim_{n \to \infty} \max_{1 \le i \le n} \frac{a_i^2}{\sum_{i=1}^n a_i^2} = 0.$$

Let $S_{a,n} = \sum_{i=1}^n a_i X_i$, $\mu_{a,n} = \mu \sum_{i=1}^n a_i$ *and* $\sigma_{a,n}^2 = \sigma^2 \sum_{i=1}^n a_i^2$. *Then*

$$Z_n = \frac{S_{a,n} - \mu_{a,n}}{\sigma_{a,n}} \xrightarrow{d} N(0,1) \text{ as } n \to \infty.$$

Proof: The Lindeberg condition holds due to the assumptions on the sequence $\{a_k\}$.

Theorem A.4.5 (Berry-Esseen): *Let* $X_1, X_2, \cdots$ *be a sequence of independent random variables with* $EX_n = 0$, $EX_n^2 = \sigma_n^2$, $s_n^2 = \sum_{i=1}^n \sigma_i^2 > 0$, $\Gamma_n^{2+\delta} = \sum_{i=1}^n E|X_i|^{2+\delta} < \infty$, *for some* $0 < \delta \le 1$ *and* $S_n = \sum_{i=1}^n X_i$. *Then there exists a universal constant* C_δ *such that*

$$\sup_{-\infty < t < \infty} \left| P\left[\frac{S_n}{s_n} \le t\right] - \Phi(t) \right| \le C_\delta \left(\frac{\Gamma_n}{s_n}\right)^{2+\delta}.$$

Remark: The above is a modification of the original theorems of [Berry (1941)] and [Esseen (1945)]. For a proof see [Chow and Teicher (1997)].

A.5   U-statistics

Results for one sample and two sample U-statistics are discussed here.

A.5.1   *One Sample U-statistics*

Let X_k, $k = 1, 2, ...$ be a sequence of independent and identically distributed random variables with common distribution function F. Let $\theta(F)$ be a functional (parameter) of the distribution F and let $\phi(X_1, X_2, \cdots, X_k)$, $k \le n$ be a statistic based on k random variables such that

$$E[\phi(X_1, X_2, \cdots, X_k)] = \theta(F)$$

and for any $l < k$ there does not exist any function $g(x_1, x_2, \cdots, x_l)$ with the property that $E[g(X_1, \cdots, X_l)] = \theta(F)$.

Then $\theta(F)$ is called an estimable functional of degree k and ϕ is called the kernel for $\theta(F)$.

Let $\phi_s(X_1, \cdots, X_k) = \frac{1}{k} \sum_p \phi(X_{i_1}, X_{i_2}, \cdots, X_{i_k})$

where $\sum_p$ is the sum over all $k!$ permutations $\{i_1, \cdots, i_k\}$ of $\{1, 2, ..., k\}$. Then ϕ_s is the symmetrized version of the kernel ϕ. Then the U-statistics corresponding to the functional $\theta(F)$ based on the kernel $\phi(X_1, \cdots, X_k)$ of degree k is

$$U_n = U(X_1, X_2, ..., X_n) = \frac{1}{\binom{n}{k}} \sum_c \phi_s(X_{i_1}, \cdots, X_{i_k})$$

where $\sum_c$ is the summation over all possible $\binom{n}{k}$ combinations $\{i_1, i_2, \cdots, i_k\}$ selected from $\{1, 2, \cdots, n\}$.

The following terms are defined in order to obtain an expression for the $Var(U_n)$. Let for $1 \leq b \leq k$,

$$\phi_b^*(x_1, \cdots, x_b) = E(\phi_s(x_1, \cdots, x_b, X_{b+1}, \cdots, X_k)),$$

and

$$\zeta_b(F) = E[(\phi_b^*(X_1, \cdots, X_b))^2] - [\theta(F)]^2.$$

Since the $X_i's$ are i.i.d., it follows that

$$\phi_b^*(X_1, \cdots, X_b) = E[\phi_s(X_1, \cdots, X_k)|X_1, \cdots, X_b]$$

and

$$\zeta_b(F) = E[\phi_s(X_1, \cdots, X_b, X_{b+1}, \cdots, X_k)\phi_s(X_1, \cdots, X_b, X_{k+1}, \cdots, X_{2k-b})]$$
$$- (\theta(F))^2$$
$$= Cov(\phi_s(X_1, \cdots, X_b, X_{b+1}, \cdots, X_k), \phi_s(X_1, \cdots, X_b, X_{k+1}, \cdots, X_{2k-b})).$$

Let $\zeta_0(F) = 0$. Then,

$$Var(U_n)$$

$$= \frac{1}{\binom{n}{k}^2} \sum_c \sum_{c\prime} E[\phi_s(X_{i_1}, \cdots, X_{i_k})\phi_s(X_{j_1}, \cdots, X_{j_k})] - (\theta(F))^2.$$

$$(A.5.1)$$

The summand $E[\phi_s(X_{i_1}, \cdots, X_{i_k})\phi_s(X_{j_1}, \cdots, X_{j_k})] - (\theta(F))^2$ in (A.5.1) equals 0 if the sets $\{i_1, \cdots, i_k\}$ and $\{j_1, \cdots, j_k\}$ have no elements in common, and equals $\zeta_b(F)$ if the two sets have exactly "b" elements in common, $b = 1, 2, ..., k$. The number of terms resulting in $\zeta_b(F)$ is

$$\binom{n}{k}\binom{k}{b}\binom{n-k}{k-b}.$$

This can be argued in the following way: From $\{1, 2, \cdots, n\}$ choose a subset $\{i_1, \cdots, i_k\}$ of k integers, for each such choice choose a further subset of b integers, which can be done in $\binom{k}{b}$ ways, for each such choice, choose the remaining $k - b$ integers for the other subset from the remaining $n - k$, this can be done in $\binom{n-k}{k-b}$ ways.

Thus

$$Var(U_n) = \frac{1}{\binom{n}{k}} \sum_{b=1}^{k} \binom{k}{b}\binom{n-k}{k-b} \zeta_b(F). \qquad (A.5.2)$$

Lemma A.5.1. *If $\zeta_k(F) < \infty$, then $\lim_{n\to\infty} nVar(U_n) = k^2\zeta_1(F)$.*

Proof: First note that an application of the Cauchy-Schwarz inequality gives

$$\zeta_b(F) \le \zeta_k(F) \text{ for all } b = 1, 2, ...k,$$

which implies that all the $\zeta_b's$ are finite. From (A.5.2), it can be seen that the coefficient (including the denominator $\binom{n}{k}$) of $\zeta_b(F)$ in the expression $nVar(U_n)$ contains $k - b$ factors of the form $(n - i)$ (with $-k \le i \le k$) in the numerator and $k - 1$ factors of the same type in the denominator. Thus as $n \to \infty$, the coefficient converges to zero for each $b \ne 1$ and to k^2 for $b = 1$. This completes the proof.

Theorem A.5.1 (The Hoeffding C.L.T): *If $\zeta_k(F) < \infty$ and $\zeta_1(F) > 0$, then*

$$\frac{\sqrt{n}(U_n - \theta(F))}{k\sqrt{\zeta_1(F)}} \xrightarrow{d} N(0,1) \text{ as } n \to \infty.$$

Proof: We will first prove the following result.

$$(i) \ \left|\sqrt{n}(U_n - \theta(F)) - \frac{k}{\sqrt{n}} \sum_{i=1}^{n}(\phi_1^*(X_i) - \theta(F))\right| = o_p(1).$$

Note that $\{\phi_1^*(X_i)\}$ is a sequence of i.i.d. random variables with mean $\theta(F)$ and variance $\zeta_1(F)$. Hence

$$E[|\sqrt{n}(U_n - \theta(F)) - \frac{k}{\sqrt{n}} \sum_{i=1}^n (\phi_1^*(X_i) - \theta(F))|^2]$$

$$= nVar(U_n) + k^2\zeta_1(F) - 2kE[\sum_{i=1}^n (U_n - \theta(F))(\phi_1^*(X_i) - \theta(F))].$$

But for each i, we have

$$E[(U_n - \theta(F))(\phi_1^*(X_i) - \theta(F))] = \binom{n-1}{k-1} / \binom{n}{k} \zeta_1(F).$$

Using Lemma (A.5.1) we obtain (i).

The Lindeberg Lévy central limit theorem can be used to obtain

$$\frac{1}{\sqrt{n}} \sum_{i=1}^n (\phi_1^*(X_i) - \theta(F)) \xrightarrow{d} N(0, \zeta_1(F)) \text{ as } n \to \infty.$$

The result then follows from the condition (i) above and Theorem (A.3.1).

A.5.2 *Two Sample U-statistics*

In nonparametric applications the more common situations where the U-statistic construction and asymptotic distribution is used occur in two sample situations.

Let $X_1, \cdots, X_{n_1}$ and $Y_1, \cdots, Y_{n_2}$ be two random samples from c.d.f.'s $F(x)$ and $G(x)$, respectively. The null hypothesis of interest is

$$H_0 : F(x) = G(x) \ \forall \ x$$

and the one sided alternative is

$$H_1 : F(x) \le G(x),$$

with strict inequality over a set of positive probability. This problem has been discussed in Chapter 7.

Then the Mann-Whitney statistic is

$$U_{n_1 n_2} = \frac{1}{n_1 n_2} \sum_{i=1}^{n_1} \sum_{j=1}^{n_2} \phi(X_i \ge Y_j),$$

where ϕ is the indicator function of the enclosed event. This is a useful statistic because

$$E[\phi(X_i \ge Y_j)] = P[X \ge Y] = \int_{-\infty}^{\infty} G(x)dF(x) = \theta(F, G), \text{ (say)},$$

which is a functional of (F, G), with the property that

$$E[\phi(X_i \geq Y_j)] = \theta(F, G), \quad \text{under } H_0 : \theta(F, G) = 1/2$$

and under $H_1 : \theta(F, G) > 1/2$.

One can generalize the ideas of one sample kernel to a two sample kernel

$$h(X_1, \cdots, X_{m_1}; Y_1, \cdots, Y_{m_2})$$

of $m_1(\leq n_1)$ and $m_2(\leq n_2)$ observations from the two random samples respectively. We call it a kernel of degree (m_1, m_2). Define the two sample U-statistics as

$$U_{n_1 n_2}(h) = \frac{1}{\binom{n_1}{m_1}\binom{n_2}{m_2}} \sum h(X_{i_1}, \cdots, X_{i_{m_1}}, Y_{j_1}, \cdots, Y_{j_{m_2}})$$

where $\{i_1, \cdots, i_{m_1}\}$ and $\{j_1, \cdots, j_{m_2}\}$ are chosen such that $1 \leq i_1 < i_2 \cdots < i_{m_1} \leq n_1$ and $1 \leq j_1 < i_2 \cdots < j_{m_2} \leq n_2$. The sum $\sum$ is over all such choices which are $\binom{n_1}{m_1}\binom{n_2}{m_2}$ in number. We then take the average of the kernels over all these choices.

It is easily seen that

$$E[U_{n_1 n_2}] = E[h(X_1, \cdots, X_{m_1}, Y_1, \cdots, Y_{m_2})] = \theta(F, G).$$

The (Wilcoxon) Mann-Whitney statistic is obviously a two sample U-statistic with degree $(1,1)$ for the parameteric function $\theta(F, G) = P[X \geq Y]$. Any testing problem which has $\theta(F, G) = 0$ under H_0 and $\theta(F, G) \neq 0$ under H_1 can then be treated with the help of the U-statistic based on a kernel which is a simple unbiased estimator of $\theta(F, G)$.

In order to carry out such tests with a specific probability of Type I error, one needs the exact and asymptotic distributions of the corresponding U-statistics. Many of these statistics are distribution free, say, based on the two sample rank statistics. If H_0 specifies that $F(x) = G(x)$ then both the random samples are from a common distribution. The two sample rank order probabilities are based on the equal probabilities for all the $(n_1 + n_2)!$ rank sets, hence are known and these will lead to distribution free tests.

However it is well known that to calculate the distribution free exact null distribution of these statistics for increasing sample sizes soon becomes unwieldy even on fast computers. Hence results proved by [Lehmann (1951)], generalizing original one sample results of [Hoeffding (1948)] are very useful. We quote the relevant two sample limit theorem here.

Let

$$V_{ij} = h(X_1, \cdots, X_i, X_{i+1}, \cdots, X_{m_1}, Y_1, \cdots, Y_j, Y_{j+1}, \cdots, Y_{m_2})$$

and

$$V'_{ij} = h(X_1, \cdots, X_i, X'_{i+1}, \cdots, X'_{m_1}, Y_1, \cdots, Y_j, Y'_{j+1}, \cdots, Y'_{m_2}),$$

where the first i $X's$ and j $Y's$ are common arguments in V_{ij} and V'_{ij} and the remaining are distinct. The $X's$ and $Y's$ are independently distributed with distributions F and G respectively.

Define

$$\rho_{ij} = Cov(V_{ij}, V'_{ij}).$$

By grouping the terms with the same number of common terms in the double summation, we get the variance of $U_{n_1 n_2}(F, G)$ as

$$Var(U_{n_1 n_2}(F, G)) = \sum_{i=1}^{m_1} \sum_{j=1}^{m_2} \frac{\binom{m_1}{i}\binom{n_1 - m_1}{m_1 - i}}{\binom{n_1}{m_1}} \frac{\binom{m_2}{j}\binom{n_2 - m_2}{m_2 - j}}{\binom{n_2}{m_2}} \rho_{ij}.$$

Let $N = n_1 + n_2$. The central limit theorem states that, if $\sigma^2_{m_1 m_2}$ is finite, $n_1/(n_1 + n_2) \to p$, $0 < p < 1$, as $n_1 + n_2 = N \to \infty$, then

$$\sqrt{N}(U_{n_1 n_2}(F, G) - \theta(F, G)) \xrightarrow{d} N(0, \zeta^2),$$

where

$$\zeta^2 = \frac{m_1^2}{p}\rho_{10} + \frac{m_2^2}{1-p}\rho_{01}.$$

The leading terms ρ_{10} and ρ_{01} are usually easy to compute.

Reverting back to the two sample Mann-Whitney statistic, we see that, under H_0

$$\rho_{10} = P[Y < X, Y' < X] - P[Y < X]^2 = \frac{1}{3} - \frac{1}{4} = \frac{1}{12},$$

and

$$\rho_{01} = 1/12.$$

Therefore,

$$\zeta^2 = \frac{1}{p}\frac{1}{12} + \frac{1}{1-p}\frac{1}{12} = \frac{1}{12p(1-p)},$$

and

$$\sqrt{N}(U - \frac{1}{2}) \xrightarrow{d} N(0, \frac{1}{12p(1-p)}) \text{ as } N \to \infty.$$

A.6   The Chernoff Savage Theorem

Let $X_1, \cdots, X_m$ and $Y_1, \cdots, Y_n$ be two independent random samples from continuous c.d.f.'s $F(x)$ and $G(x)$, respectively. Let $N = m + n$, $\lambda_N = m/N$ and $\lim_{N \to \infty} \lambda_N = \lambda$ such that $0 < \lambda_0 \le \lambda \le 1 - \lambda_0 < 1$ for some $\lambda_0 \le 1/2$. Let $F_m(x)$ and $G_n(x)$ be the empirical distribution functions corresponding to the two random samples. Then $H_N(x) = \lambda_N F_m(x) + (1 - \lambda_N)G_n(x)$ is the combined empirical distribution function based on the two random samples taken together. Let $J_N(u)$, $0 < u < 1$ be a sequence of functions such that $J_N(i/(N + 1))$, $i = 1, 2, ..., N$ provide certain scores. Define a statistic

$$T_N = \int_{-\infty}^{\infty} J_N(\frac{N}{N + 1} H_N(x)) dF_m(x).$$

This integral actually is the sum of scores $J_N(R_i/(N + 1))$ corresponding to $R_1, R_2, \cdots, R_m$, the ranks of the observations from the X sample in the combined order. Many of the sample statistics, such as the Wilcoxon statistic, the normal scores statistic, etc., are in this form. The asymptotic normality of such statistics under certain conditions was first obtained by [Chernoff and Savage (1958)]. We first state the required conditions and introduce some notation.

The Conditions
(i) $J_N(u) \to J(u)$, $0 < u < 1$ where $J(u)$ is not a constant.

(ii) $\int_{-\infty}^{\infty} \left[J_N(\frac{N}{N+1} H_N(x)) - J(\frac{N}{N+1} H_N(x)) \right] dF_m(x) = o_p(N^{-1/2})$.

(iii) $\left| \frac{d^i J(u)}{du^i} \right| \le K(u(1 - u))^{-i-1/2+\delta}$ for some $\delta > 0$ and $i = 0, 1$ with K a constant.

Let

$$H(x) = \lambda_N F(x) + (1 - \lambda_N)G(x).$$

Note that H depends on N. Further let

$$\mu_N = \int_{-\infty}^{\infty} J(H(x)) dF(x)$$

and

$$\sigma_N^2 = 2(1 - \lambda_N)\{ \int \int_{-\infty < x < y < \infty} G(x)(1 - G(y))J'(H(x))J'(H(y)) dF(x) dF(y)$$
$$+ \frac{1 - \lambda_N}{\lambda_N} \int \int_{-\infty < x < y < \infty} F(x)(1 - F(y))J'(H(x))J'(H(y)) dG(x) dG(y)\},$$

where $J'(x)$ denotes the first derivative of $J(x)$ w.r.t. x.

Choose a x_0 so that $H(x_0) = 1/2$ and define

$$B(x) = \int_{x_o}^x J'(H)dG, \quad B^*(x) = \int_{x_o}^x J'(H)dF.$$

From condition (iii) above and the fact that $dF \leq KdH$, where K is a constant, one obtains $|\mu_N| \leq K \int_\infty^\infty (H(1-H))^{\delta-1/2}dH$, for some $\delta > 0$. This proves that μ_N is finite.

Theorem A.6.1 (Chernoff-Savage): *Under the conditions (i) to (iii) and if $\sigma_N \neq 0$,*

$$\lim_{N \to \infty} P\left(\frac{\sqrt{n}(T_N - \mu_N)}{\sigma_N} \leq t\right) = \Phi(t).$$

Under the additional condition that $B(X_1)$ and $B^*(Y_1)$ have variances bounded away from zero with respect to (F, G, λ_N), the above convergence to the standard normal distribution is uniform in F, G, and λ_N.

Remark A.6.1. [Chernoff and Savage (1958)] had assumed a second derivative condition on J but this condition was subsequently removed by [Govindarajulu *et al.* (1967)] and [Pyke and Shorack (1968)] using proofs based on properties of the empirical process. A proof is also given by [Puri *et al.* (1971)] and [Akritas (1984)].

Below we give a sketch of the proof.
Proof of the Chernoff-Savage theorem: Let

$$T_N^* = \int_{-\infty}^\infty J(\frac{N}{N+1}H_N(x))dF_m(x).$$

From condition (ii) above it follows that

$$T_N - T_N^* = o_p(N^{-1/2}).$$

Thus by Theorem (A.3.1) it is enough to obtain the asymptotic distribution of the statistic $\frac{\sqrt{N}(T_N^* - \mu_N)}{\sigma_N}$.
We can write

$$J(\frac{N}{N+1}H_N) = J(H) + (H_N - H)J'(H) - \frac{H_N}{N+1}J'(H)$$

$$+\{J(\frac{N}{N+1}H_N) - J(H)(\frac{N}{N+1}H_N - H)J'(H)\}.$$

Thus

$$\sqrt{N}(T_N^* - \mu_N)$$

$$= \sqrt{N} \int J(H)d(F_m - F) + \sqrt{N} \int (H_N - H)J'(H)dF$$

$$- \frac{\sqrt{N}}{N+1} \int H_N J'(H)dF_m + \sqrt{N} \int (H_N - H)J'(H)d(F_m - F)$$

$$+ \sqrt{N} \int \{J(\frac{N}{N+1}H_N) - J(H) - (\frac{N}{N+1}H_N - H)J'(H)\}dF_m$$

$$= B_{1N} + B_{2N} + C_{1N} + C_{2N} + C_{3N}, \text{ say.}$$

Below we proceed to show that $(B_{1N} + B_{2N})/\sigma_N$ converges in distribution to a standard normal variable.

Note that, using integration by parts,

$$B_{2N} = \sqrt{N} \int (H_N - H)J'(H)dF$$

$$= \sqrt{N} \int (H_N - H)dB^*$$

$$= -\sqrt{N} \int B^* d(H_n - H). \tag{A.6.3}$$

Using this,

$$\lambda_N B^*(x) + (1 - \lambda_N)B(x) = J(H(x)) - J(H(x_0))$$

and the fact that $\int d(F_m - F) = 0$, we obtain

$$B_{1N} + B_{2N} = \sqrt{N}(1 - \lambda_N)\{\int Bd(F_m - F) - \int B^* d(G_n - G)\}$$

$$= \sqrt{N}(1 - \lambda_N)\{\frac{1}{m}\sum_{i=1}^{m}(B(X_i) - E[B(X_1)])$$

$$- \frac{1}{n}\sum_{i=1}^{n}(B^*(Y_i) - E[B(Y_1)])\}. \tag{A.6.4}$$

The equation (A.6.4) is a sum of independent random variables with mean zero and the sum of the variances

$$s_N^2 = N(1 - \lambda_N)^2\{\frac{1}{m}Var(B(X_1)) + \frac{1}{n}Var(B^*(Y_1))\}. \tag{A.6.5}$$

In our notation $F_1(x) = I[X_1 \leq x]$ and thus

$$B(X_1) = \int B(x)dF_1.$$

Using integration by parts, we can write

$$B(X_1) - E[B(X_1)] = - \int (F_1 - F)J'(H)dG.$$

Now condition (iii), and the fact that $dG \leq KdH$ implies that

$$E|B(X_1) - E[B(X_1)]|^{2+\delta'} \leq K \int (H(1-H)^{(-1/2+\delta)(2+\delta')}dH, \quad (A.6.6)$$

where K is a generic constant. The bound above is finite for a $\delta' > 0$ satisfying $(-12 + \delta)(2 + \delta') > -1$. Further

$$Var(B(X_1))$$

$$= E[\int (F_1(x) - F(x))J'(H(x))dG(x) \int (F_1(y) - F(y))J'(H(y))dG(y)]$$

$$+ 2 \int \int_{-\infty < x < y < \infty} F(x)(1 - F(y))J'(H(x))J'(H(y))dG(x)dG(y),$$

using the Fubini theorem and the symmetry of the integral in x and y

$$E[(F_1(x) - F(x))(F_1(y) - F(y))] = F(\min\{x, y\}) - F(x)F(y).$$

Similar arguments lead to the finiteness of $E|B^*(Y_1) - E[B^*(Y_1)]|^{2+\delta'}$ and to

$$Var(B^*(Y_1))$$

$$= 2 \int \int_{-\infty < x < y < \infty} G(x)(1 - G(y))J'(H(x))J'(H(y))dF(x)dF(y).$$

Now consider the N independent random variables

$$\sqrt{N}(1 - \lambda_N)(B(X_i) - E[B(X_1)])/m, \sqrt{N}(1 - \lambda_N)(B^*(Y_j) - E[B(Y_1)])/n;$$

$i = 1, \cdots, m; j = 1, \cdots, n$. The sum of their $(2+\delta')$-th absolute moments is bounded above by $KN^{(2+\delta')/2}(1-\lambda_n)^{2+\delta'}(1/m^{1+\delta'} + 1/n^{1+\delta'})$, which tends to zero as $N \to \infty$ since $m/N \to \lambda$, $n/N \to (1 - \lambda)$ and m and n tend to ∞. Also, $H = \lambda_N F(x) + (1 - \lambda_N)G(x) \to \lambda F + (1 - \lambda)G$. Further, from condition (iii) and the dominated convergence theorem, s_N^2 converges to a finite quantity as $N \to \infty$. Thus by the Liapounov Theorem, we obtain that

$$\frac{B_{1N} + B_{2N}}{s_N} \xrightarrow{d} N(0, 1).$$

But from (A.6.3), (A.6.5) and the expressions for $Var(B(X_1))$ and $Var(B^*(Y_1))$ we see that $s_N = \sigma_N$.

The remaining terms C_{1N}, C_{2N} and C_{3N} above are $o_p(1)$ uniformly in (F, G, λ_N). For a proof of this we refer to [Chernoff and Savage (1958)] and [Puri *et al.* (1971)].

To see that the convergence to the standard normal is uniform in (F, G, λ_N), we further note that the properties of the empirical distribution used, depend only on the properties of a empirical distribution function corresponding to a random sample from a distribution that is Uniform over $(0, 1)$. Hence without loss of generality H can be assumed to be a $U(0, 1)$ distribution. To see the uniform convergence of $B_{1N} + B_{2N}$ to the standard normal, we note that the bound in the Berry-Esseen theorem (Theorem A.4.5) depends only on the variances and the $(2+\delta')$-th moments of $B(X_1)$ and $B^*(Y_1)$. From (A.6.6), it can be seen that the $(2+\delta)$-th moments are bounded by $\int_0^1 (u(1-u))^{2+\delta'} du$ and thus are independent of (F, G, λ_N). Since the variances are given to be bounded away from zero, the bound does not depend on (F, G, λ_N).

A.7   Linear Functions of Order Statistics

For tests of exponentiality and some other applications we need the asymptotic distribution of linear functions of order statistics. The following theorem provides the asymptotic normality under certain conditions.

Theorem A.7.1. *[Moore* et al. *(1968)]: Let* $X_{(1)} < X_{(2)} < \cdots < X_{(n)}$ *be the order statistics corresponding to a random sample from c.d.f.* $F(x)$. *Let*

$$T_n = \frac{1}{n} \sum_{i=1}^{n} L(\frac{i}{n}) X_i,$$

where $L(u)$ *is a function on [0,1]. Suppose* $L(u)$ *and* $F(x)$ *satisfy the following conditions.*
(F1): *The c.d.f.* $F(x)$ *is continuous.*
(F2): $E[|X|] < \infty$.
(W): *The derivative* $L'(u)$ *of* $L(u)$ *is continuous and of bounded variation on [0,1].*

Further let

$$\mu = \int_{-\infty}^{\infty} x L(F(x)) dF(x)$$

and

$$\sigma^2 = 2 \int\int_{-\infty < s < t < \infty} L(F(s))L(F(t))F(s)(1 - F(t))dsdt.$$

Then

$$P\left[\frac{\sqrt{n}(T_n - \mu)}{\sigma} \le t\right] \to \Phi(t) \ \ as \ n \to \infty$$

provided $0 < \sigma^2 < \infty$.

Proof: Since $F(x)$ is continuous, $F(X_1)$ is a uniform random variable over $[0, 1]$. Let U_n denote the empirical distribution function of the $U(0, 1)$ random variables $F(X_1), \cdots, F(X_n)$ and define the inverse F^{-1} of F as

$$F^{-1}(t) = inf\{x : t \le F(x)\}.$$

By the above relations, and the change of variable and the integration by parts formulae,

$$\begin{aligned}
T_n - \mu_n &= \int_{-\infty}^{\infty} xL(F_n(x))dF_n(x) - \int_{-\infty}^{\infty} xL(F(x))dF(x) \\
&= \int_0^1 F_n^{-1}(u)L(u)du - \int_0^1 F^{-1}(u)L(u)du \\
&= \int_0^1 F^{-1}(U_n^{-1}(u))L(u)du - \int_0^1 F^{-1}(u)L(u)du \\
&= \int_0^1 F^{-1}(u)L(U_n(u))dU_n(u) - \int_0^1 F^{-1}(u)L(u)du.
\end{aligned}$$

Let $W_n(u) = \sqrt{n}(U_n(u) - u)$. From the mean value theorem

$$L(U_n(u)) - L(u) = L'(V_n(u))(U_n(u) - u),$$

where $|V_n(u) - u| < |U_n(u) - u|$.
Therefore we may write

$$\sqrt{n}(T_n - \mu_n) = B_{1n} + R_{1n} + R_{2n},$$

where

$$B_{1n} = \int_0^1 F^{-1}(u)L'(u)W_n(u)du + \int_0^1 F^{-1}(u)L(u)dW_n du,$$

$$R_{1n} = \int_0^1 F^{-1}(u)\{L'(V_n(u)) - L'(u)\}W_n(u)dU_n(u),$$

$$R_{2n} = n^{-1/2}\int_0^1 F^{-1}(u)L'(u)W_n(u)dW_n(u).$$

Since the assumption $(F2)$ implies that

$$\lim_{u \to 0^+} u F^{-1}(u) = \lim_{u \to 1^-} (1-u)F^{-1}(u) = 0,$$

we can integrate by parts to obtain

$$B_{1n} = \int_0^1 L(u)W_n(u)dF^{-1}(u).$$

Now the integral is $\sqrt{n}$ times the average of n i.i.d. random variables

$$\{\int_0^1 (I[F(X_i) \le u] - u)L(u)dF^{-1}(u), \ i = 1, ..., n\}$$

with common mean 0. Since σ^2 is finite, the common variance can be obtained using the Fubini Theorem as follows:

$$Var(\int_0^1 (I[F(X_i) \le u] - u)L(u)dF^{-1}(u))$$

$$= E \int_0^1 \int_0^1 (I[F(X_1) \le u] - u)L(u)(I[F(X_1) \le v] - v)L(v)dF^{-1}(v)dF^{-1}(u)]$$

$$= \int_0^1 \int_0^1 (\min(u,v) - uv)dF^{-1}(u)dF^{-1}(v)$$

which by the change of variable formula and the symmetry of the integrals in u and v reduces to σ^2. From the Lindberg-Lévy theorem, it follows that

$$B_{1n} \overset{d}{\to} N(0,1) \ \text{ as } n \to \infty.$$

We, next show that the remaining two terms R_{1n} and R_{2n} are $o_p(1)$. Note that

$$|R_{1n}| \le \|L'(V_n) - L'\|\|W_n\|\int_0^1 |F^{-1}(u)|dU_n(u),$$

where $\|.\|$ denotes the supremum norm. By the Glivenko-Cantelli theorem and the uniform continuity of $L'(u)$, the first factor of the bound above converges to zero a.s. A theorem due to Kolmogorov (see [Billingsley (1995)], p. 104) asserts that the second factor converges in distribution to a proper random variable. The SLLN asserts that the third factor converges to $\int_0^1 |G(u)|du$ which is finite due to assumption $(F2)$. This gives $|R_{1n}| = o_p(1)$.

The term R_{2n} can be expressed as a sum of integrals over the continuity set of W_n and its complement as follows:

$$R_{2n} = \frac{1}{2}n^{-1/2} \int_0^1 F^{-1}(u)L'(u)d[W_n(u)]^2$$

$$+ 1/2n^{-3/2} \sum_{i=1}^n F^{-1}(F(X_i))L'(F(X_i)), \text{ a.s.}$$

The second term on the right converges to 0 by the SLLN due to the assumptions on $F(x)$ and $L(u)$. After integration by parts the first term equals

$$\frac{1}{2}n^{-1/2} \int_0^1 [W_n(u)]^2 d[F^{-1}(u)L'(u)].$$

For a and b in $[0,1]$ $(a < b)$, with $F^{-1}(a)$ and $F^{-1}(b)$ finite, the term

$$n^{-1/2} \int_a^b [W_n(u)]^2 d[F^{-1}(u)L'(u)],$$

converges to zero in probability because $\|W_n\|$ is $O_p(1)$ and the integral involved is finite. Thus we need to show that the integrals over $[0,a]$ and $[b,1]$ converge to zero in probability. We can assume b to be a continuity point of $F^{-1}(u)$ and need only consider the case $F^{-1}(1) = \infty$. Let $V(u)$ denote the total variation of $F^{-1}L'(u)$ on $[b,1]$, then

$$E|n^{-1/2} \int_b^1 [W_n(u)]^2 d[F^{-1}(u)L'(u)]|$$

$$\leq E[n^{-1/2} \int_b^1 [W_n(u)]^2 dV(u)]$$

$$= n^{-1/2} \int_b^1 u(1-u)dV(u).$$

Thus it is enough to show that

$$\int_b^1 u(1-u)dV(u) < \infty.$$

Let $V_s^t(f)$ denote the total variation of f on $[s,t]$. Then

$$V(u) \leq \sup_{[b,u]} |F^{-1}(t)|V_b^u(L') + \sup_{[b,u]} |L'(t)|V_b^u(F^{-1})$$

$$= F^{-1}(u)V_b^u(L') + \sup_{[b,u]} |L'(t)|[(F^{-1}(u) - F^{-1}(b)]$$

$$\leq KF^{-1}(u),$$

for some constant $K > 0$. Integrating $\int_b^1 u(1-u)dV(u)$ by parts and using the above bound for $V(u)$ establishes the finiteness of the integral. Convergence to 0 in probability of the integral over $[0, a]$ can be shown by a similar argument.

The asymptotic normality of $\frac{\sqrt{n}(T_n - \mu)}{\sigma}$ now follows from the above results and the Theorem (A.3.1). $\square$

Remark A.7.1. [Stigler (1974)] has obtained the asymptotic normality for a general c.d.f. $F(x)$ under the assumptions
(F'): $E[X_1^2] < \infty$.
(W'): $L(u)$ is bounded and continuous a.e. F^{-1} on $(0, 1)$.

Remark A.7.2. Asymptotic normality of the statistic $\frac{1}{n}\sum_{i=1}^n L(\frac{i}{n})h(X_i)$ known as L-statistic, under certain conditions on the weight function L and the function h, has been obtained by many authors. References and the various conditions needed may be found in the book by ([Shorack and Wellner (1986)], Chapter 19).

A.8 A Martingale Central Limit Theorem

Certain test statistics can be expressed as stochastic integrals with respect to counting processes which in turn involve integrals w.r.t. innovation martingales. Thus martingale central limit theorems may be used in obtaining the asymptotic distribution of these statistics. Below we state the central limit theorem used in this book. More details can be found in [Fleming and Harrington (1991)], [Karr (1991)], [Andersen *et al.* (1993)].
  We first state some definitions. Let $(\Omega, \mathcal{F}, P)$ be a probability space.

Definition A.8.1. A family of sub σ-fields $\{\mathcal{F}_t, t \geq 0\}$ of $\mathcal{F}$ is called a *filtration* if $s \leq t$ implies $\mathcal{F}_s \subset \mathcal{F}_t$.

Definition A.8.2. A *counting process* is a stochastic process $\{N(t), t \geq 0\}$ with $N(0) = 0$, $N(t) < \infty$ a.s., and whose paths are piecewise constant, right-continuous and have only jump discontinuities with jump size $+1$.

Definition A.8.3. A stochastic process $\mathbf{X} = \{X(t), t \geq 0\}$ is said to be *adapted* to a filtration $\{\mathcal{F}_t, t \geq 0\}$ or $\mathcal{F}_t$-adapted if for every t, $X(t)$ is $\mathcal{F}_t$-measurable.

Definition A.8.4. A stochastic process $\mathbf{X} = \{X(t), t \geq 0\}$ is called a *martingale* with respect to a filtration $\{\mathcal{F}_t, t \geq 0\}$ or $\mathcal{F}_t$-martingale if:
1. $\mathbf{X}$ is adapted to a filtration $\{\mathcal{F}_t, t \geq 0\}$,
2. $E[|X(t)|] < \infty$ for all t,
3. $E[X(t)|\mathcal{F}_u, u \leq s] = X(s)$ a.s. for all $0 \leq s \leq t$.

Definition A.8.5. A stochastic process $\mathbf{X} = \{X(t), t \geq 0\}$ is said to be *predictable* with respect to a filtration $\{\mathcal{F}_t, t \geq 0\}$ or $\mathcal{F}_t$-predictable, if for each $t > 0$, $X(t)$ is $\mathcal{F}_{t-}$-measurable where $\mathcal{F}_{t-} = \sigma\{\bigcup_{u<t} \mathcal{F}_u\}$, (i.e. the smallest σ-field generated by the collection $\bigcup_{u<t} \mathcal{F}_u$).

Definition A.8.6. A nonnegative random variable T is called a *stopping time* with respect to a filtration $\{\mathcal{F}_t, t \geq 0\}$ if $[T \leq t] \in \mathcal{F}_t$ for all $t \geq 0$.

Definition A.8.7. A property is said to hold *locally* for a stochastic process $\mathbf{X}$ if the property is satisfied by the stopped process $\{X(t \wedge T_n); t \geq 0\}$ for each n, where $\{T_n, n = 1, 2, \cdots\}$ is an increasing sequence of stopping times with $\lim_{n \to \infty} T_n = \infty$.

From the Doob-Meyer decomposition we have that if $\mathbf{N} = \{N(t), t \geq 0\}$ is a counting process adapted to a right continuous filtration $\{\mathcal{F}_t, t \geq 0\}$ with $EN(t) < \infty$ for all t, then there exists a unique increasing right continuous $\mathcal{F}_t$-predictable process $\mathbf{\Lambda} = \{\Lambda(t), t \geq 0\}$ such that $\Lambda(0) = 0$ a.s., $E\Lambda(t) < \infty$ for all t, and $\{M(t) = N(t) - \Lambda(t), t \geq 0\}$ is a right continuous $\mathcal{F}_t$-martingale called the innovation martingale of the process $\mathbf{N}$. The process $\mathbf{\Lambda}$ is called the compensator of the counting process with respect to the filtration $\{\mathcal{F}_t\}$.

The following inequality is used to prove strong consistency of the Kaplan-Meier and Nelson-Aalen estimators.

Theorem A.8.1. *(Lenglart's Inequality) Let $\mathbf{X}$ be a right-continuous $\mathcal{F}_t$-adapted process, and $\mathbf{Y}$ a non-decreasing predictable process with $Y(0) = 0$. Suppose, for all bounded stopping times T, $E[X(T)] \leq E[Y(T)]$. Then for any stopping time T, and every $\epsilon, \eta > 0$,*

$$P\left[\sup_{t \leq T} |X(t)| \geq \epsilon\right] \leq \frac{\eta}{\epsilon} + P[Y(T) \geq \eta].$$

A proof of the above theorem is in ([Shorack and Wellner (1986)], Inequality B.4.1).

Corollary A.8.1. *Let* $\mathbf{M} = \mathbf{N} - \mathbf{\Lambda}$ *be the martingale corresponding to the counting process* $\mathbf{N}$ *with compensator* $\mathbf{\Lambda}$, *which is locally square integrable. Suppose* $\mathbf{Y}$ *is a predictable and locally bounded process with respect to the same filtration. Then for any stopping time* T *such that* $P[T < \infty] = 1$, *and every* ϵ, $\eta > 0$,

$$P\left[\sup_{t \leq T}\{\int_0^t Y(s)dM(s)\}^2 \geq \epsilon\right] \leq \frac{\eta}{\epsilon} + P\left[\int_0^T Y^2(s)d\Lambda(s) \geq \eta\right].$$

For a proof see ([Fleming and Harrington (1991)], pp. 11).

We next state a central limit theorem involving stochastic integrals with respect to the innovation martingales. Suppose for each n and $i = 1, \cdots, k_n$, $\mathbf{N}_i^{(n)}$ is counting process with continuous compensator $\mathbf{\Lambda}_i^{(n)}$ and innovation martingale $\mathbf{M}_i^{(n)}$, and is adapted to a right continuous filtration $\{\mathcal{F}_t, t \geq 0\}$. Further, for each n and $i = 1, \cdots, n$, suppose $\{Y_i^{(n)}(s), s \geq 0\}$ is a locally bounded $\mathcal{F}_t$−predictable process. Here $k_n = k$ can be fixed or can depend on n, for example, $k_n = n$.

Theorem A.8.2. *If for every* $t \leq \tau$, *as* $n \to \infty$

$$\sum_{i=1}^{k_n} \int_0^t (Y_i^{(n)}(s))^2 d\Lambda_i^{(n)}(s) \xrightarrow{P} V(t), \qquad (A.8.7)$$

where $V(t)$ *is a non-random function and for all* $\epsilon > 0$,

$$\sum_{i=1}^{k_n} \int_0^t (Y_i^{(n)}(s))^2 I_{[|Y_i^{(n)}(s)| \geq \epsilon]} d\Lambda_i^{(n)}(s) \xrightarrow{P} 0, \qquad (A.8.8)$$

then, as $n \to \infty$,

$$\sum_{i=1}^{n} \int_0^t Y_i^{(n)}(s) dM_i^{(n)}(s)$$

converges weakly in the Skorohod topology to a mean zero Gaussian martingale on $D[0, \tau]$, *where* $D[0, \tau]$ *denotes the space of functions on* $[0, \tau]$ *which are right continuous with left-hand limits.*

Proof can be found in ([Fleming and Harrington (1991)], Chapter 5).

If for each i, $\Lambda_i^{(n)}(t) = \int_0^t \lambda_i^{(n)}(s)ds$ and $V(t) = \int_0^t v(s)ds$ then A.8.7 holds if as $n \to \infty$,

$$\sum_{i=1}^{k_n} (Y_i^{(n)}(s))^2 \lambda_i^{(n)}(s) \xrightarrow{P} v(s) > 0,$$

for all $s \in [0, \tau]$.

If $k_n = k$ is fixed then A.8.8 holds if as $n \to \infty$,

$$Y_i^{(n)}(s) \xrightarrow{P} 0$$

for all $j = 1, ..., k$ and $s \in [0, \tau]$.

A.9 Asymptotic Distribution of the Hellinger Distance Based Goodness-of-Fit-Statistic

We discussed a goodness-of-fit test based on the Hellinger distance in Chapter 10. Below we list the assumptions required for the asymptotic normality of the statistic.

Assumptions: [Beran (1977)].

A1. The kernel density $K(\cdot)$ is symmetric about zero, has compact support and is twice continuously differentiable.

A2. The parameter space Θ is a compact subset of $R^p, \theta_1 \neq \theta_2$ implies $f_{\theta_1} \neq f_{\theta_2}$ on a set of positive Lebesgue measure and for almost every $x, f_\theta(x)$ is continuous in θ.

A3. For every $\theta \in \Theta, f_\theta(x)$ is separated and positive on a compact interval I and is twice continuously differentiable (in x).

Let $h_t = f_t^{1/2}$. For every $\theta \in$ interior (Θ), the following assumptions hold.

A4. For every $x \notin N$ (a Lebesgue null set) and for every t in some neighbourhood of $\theta, h_t(x)$ has first order partial derivatives $\{\dot{h}_t^{(j)}(x), 1 \leq j \leq p\}$ with respect to t which are continuous in t at $t = \theta$. For every $j, \int (\dot{h}_t^{(j)}(x))^2 dx$ is continuous in t at $t = \theta$.

A5. For every $x \notin N$ (a Lebesgue null set) and for every t in some neighbourhood of $\theta, h_t(x)$ has second partial derivatives $\{\ddot{h}_t^{(j,k)}(x), 1 \leq j, k \leq p\}$ with respect to t which are continuous in t at $t = \theta$. Further for every $(j, k), \int (\ddot{h}_t^{(j,k)}(x))^2 dx$ is continuous at $t = \theta$.

A6. The matrix $\langle \int \ddot{h}_t^{(j,k)}(x) f_t^{1/2}(x) dx \rangle$ is non-singular.

A7. For every $t \in \Theta, h_t(x)$ is continuous in x.

A8. $\lim_{n \to \infty} nC_n^{7/2} = 0$ and $\lim_{n \to \infty} nC_n^3 = \infty$.

A9. There exists a positive finite constant s depending on f_θ such that $\sqrt{n}((S_n - s)$ is $O_p(1)$ under f_θ.

We note that under the above assumptions, the limiting distribution of $\sqrt{n}(\hat{\theta}_n - \theta)$ is normal with mean 0 and variance $\frac{1}{4}[\int \dot{h}_\theta(x)\dot{h}_\theta^T(x)dx]^{-1}$ under f_θ, where $\dot{h}_\theta(x) = (\dot{h}_\theta^{(1)}(x), \cdots, \dot{h}_\theta^{(p)}(x))^T$ with T denoting the transpose.

The result holds for parametric families $\{f_\theta, \theta \in \Theta\}$ where Θ is not compact but can be embedded within a compact set. This, for example, is possible for a location scale family $\{\sigma^{-1}f(\sigma^{-1}(x-\mu)); \sigma > 0, -\infty < \mu < \infty\}$ where f is continuous by the transformation $\mu = \tan\theta_1, \sigma = \tan\theta_2$

$$\theta = (\theta_1, \theta_2) \in (-\pi/2, \pi/2) X(0, \pi/2) = \Theta'$$

$$f_\theta(x) = (\tan\theta_2)^{-1}[f((\tan\ \theta_2)^{-1}(x - tan\ \theta_1))],$$

$\theta \in \Theta'$ and $f_\theta^{1/2} - g_t^{1/2}$ can be extended to a continuous function on $\overline{\Theta} = [-\pi/2, \pi/2] \times [0, \pi_2]$ which is compact.

A.10   Asymptotic Relative Efficiency

The classical 'goodness' criterion for a test of hypothesis is its power at a fixed alternative, or the value of its power function at various points in the alternative hypothesis. This works well in a finite dimensional alternative hypothesis but not tractable in the infinite dimensional case. Pitman proposed a modified measure of goodness for comparison of two tests. The idea was to try and see how many observations the two tests of same size require for attaining the same power for the same fixed alternative.

Then, use the ratio of the required sample sizes as the measure of relative efficiency, the test that required the smaller number of observations was the more efficient.

This approach too attracts the same criticism that there are too many points in the alternative and the values may be different for different fixed values of the power. Practically all tests that we look at will (or should) have the property of 'consistency', meaning the power should tend to 1 as the sample size increases.

Pitman's approach was to fix a parametric family of alternatives and consider a sequence

$$\theta_n = \theta_0 + n^{-\frac{1}{2}}\delta + o(n^{-\frac{1}{2}}), \ \delta > 0 \qquad\qquad (A.10.9)$$

where θ_0 gives a member of the null hypothesis and θ_n falls in the alternative. For this sequence of alternatives, [Noether (1955)] considered test statistics T_n which have asymptotically normal distribution, i.e., there are sequences $\mu_n(\theta)$ and $\sigma_n(\theta)$ such that the standardized statistic

$$\frac{T_n - \mu_n(\theta)}{\sigma_n(\theta)} \ \xrightarrow{d}\ N(0,1) \ \text{ as } \ n \to \infty.$$

Assume that

(i) $\mu_n(\theta)$ is differentiable at θ_0 and $\mu'_n(\theta_0) > 0$,

(ii) $\lim_{n\to\infty} \frac{\mu_n(\theta_n)}{\mu_n(\theta_0)} = \lim_{n\to\infty} \frac{\sigma_n(\theta_n)}{\sigma_n(\theta_0)} = 1$

and

(iii) $\lim_{n\to\infty} n^{-\frac{1}{2}} \frac{\mu'_n(\theta_0)}{\sigma_n(\theta_0)} = c > 0$.

Then, Noether proved that the power of the one-sided test based on T_n which rejects for large values of the statistics, tends to $\Phi(c\delta + z_\alpha)$ as $n \to \infty$ where z_α is the αth quantile of the standard normal distribution. Thus, the limiting power is in terms of δ which specifies the alternative and c, which is termed the 'efficacy' of the test. The Pitman-Noether asymptotic relative efficiency (ARE) of a test 1 with respect to test 2 is defined as

$$e_{12} = (\frac{c_1}{c_2})^2. \qquad (A.10.10)$$

One can see that e_{12} is the $\lim \frac{n_2}{n_1}$ as both n_1 and n_2 tend to ∞ such that the test will have the same limiting power for the same alternative given by θ_n.

In a way one may say that the slopes of the two power functions at θ_0 are being compared and the one which has the greater slope is termed to be the more efficient with e_{12} providing a precise numerical value for the comparison.

Example: Consider the one sample location problem where $X_1, X_2, \ldots, X_n$ are i.i.d from $F_\theta(x) = F(x - \theta)$. Let σ^2 be the variance of X_1. Three competing tests for testing $H_0 : \theta = \theta_0 = 0$ against $H_1 : \theta > 0$ are based on

(i) $T_{1n} = \bar{X}$.

(ii) $T_{2n} = \frac{1}{n} \sum_{i=1}^{n} \Psi(X_i)$, where $\Psi(X_i) = I(X_i > 0)$.

(iii) $T_{3n} = \sum_{i=1}^{n} \Psi(X_i) R_i^+$, where R_i^+ is the rank of $|X_i|$ in $|X_1|, |X_2|, \ldots, |X_n|$. Then T_{3n} can be written as follows

$$T_{3n} = nT_{2n} + \binom{n}{2} T_{4n},$$

where $T_{4n} = \frac{1}{\binom{n}{2}} \sum_{1 \leq i < j \leq n} I(X_i + X_j > 0)$.

Then it is easy to see that
$\mu_{1n}(\theta) = \theta$, $\sigma_{1n}^2(\theta_0) = \frac{\sigma^2}{n}$, $c_1 = \frac{1}{\sigma}$,

$\mu_{2n}(\theta) = 1 - F(-\theta)$, $\sigma_{2n}^2(\theta_0) = \frac{1}{4n}$, $c_2 = 2f(0)$,

$\mu_{3n}(\theta) = \frac{2}{n-1}[1 - F(-\theta)] + \int_{-\infty}^{\infty} [1 - F(-x - \theta)] dF(x - \theta)$

$\sigma_{3n}^2(\theta_0) = \frac{1}{3n}, \quad c_3 = 2\sqrt{3} \int_{-\infty}^{\infty} f^2(x)dx.$

Then the efficiency of T_{1n} with respect to T_{2n} and T_{3n} is given by

$$e_{12} = 4f^2(0)\sigma^2, \quad e_{13} = 12\sigma^2 \left[\int_{-\infty}^{\infty} f^2(x)dx \right]^2. \qquad (A.10.11)$$

When, underlying distribution is $N(0, \sigma^2)$, we have

$$e_{12} = 0.636, \quad e_{13} = 0.954.$$

Hence, the sign test with level α needs 100 observations to achieve a given power, the classical t-test would need 65 provided the underlying distribution is $N(0, \sigma^2)$. On the other hand if t-test needs 95 observations the Wilcoxon signed rank statistic needs 100 observations. In particular one can see that for any arbitrary continuous and symmetric distribution $F(x)$, we have from [Lehmamm (1975)]

$$e_{13} \geq .864. \qquad (A.10.12)$$

For the two sample location problem, one can check that the efficiency of the two-sample t-test (T_{5n}) (the optimal test when the underlying distribution is normal) with respect to the Wilcoxon-Mann-Whitney test (T_{6n})

$$e_{56} = 12\sigma^2 \left[\int_{-\infty}^{\infty} f^2(x)dx \right]^2, \qquad (A.10.13)$$

where σ^2 is the variance of the parent distribution. Also, a lower bound for the efficiency of the two-sample t-test (T_{5n}) with respect to the Normal-Scores test (T_{7n}) is given below

$$e_{57} \geq 1. \qquad (A.10.14)$$

Thus, the Normal scores test will be more efficient than the two sample t-test for any distribution other than the normal.

Comments:

(i) In the above case the test statistic of interest has limiting normal distribution. However, if the two test statistics T_{1n} and T_{2n} have limiting non-central chi-square distribution with the same degrees of freedom, then for testing against a sequence of alternatives converging to the null hypothesis, the Pitman ARE of T_{1n} with respect to T_{2n} is the ratio of the two noncentrality parameters (see, for example, [Andrews (1954)], [Hannan (1956)] and [Puri *et al.* (1971)]).

(ii) If T_{1n} has limiting normal distribution and T_{2n} have limiting non-central chi-square distribution with 2 degrees of freedom one can use the method proposed by [Rothe (1981)] to compare the two tests. See [Singh (1984)] for a review on concepts of asymptotic efficiency.

BIBLIOGRAPHY

Aalen, O. O. (1975). *Statistical inference for a family of counting processes* (Institute of Mathematical Statistics, University of Copenhagen).

Akritas, M. G. (1984). A simple proof for the chernoff-savage theorem, *Statistics & probability letters* **2**, 1, pp. 39–44.

Andersen, P. K., Borgan, O., Gill, R. D., and Keiding, N. (1993). *Statistical Models Based on Counting Processes* (Springer Science & Business Media).

Andrews, F. C. (1954). Asymptotic behavior of some rank tests for analysis of variance, *The Annals of Mathematical Statistics*, pp. 724–736.

Arnold, B. C., Balakrishnan, N., and Nagaraja, H. N. (1992). *A first course in order statistics*, Vol. 54 (Siam).

Barlow, R. E. and Proschan, F. (1981). *Statistical Theory or Reliability and Life Testing: Probability Models* (To Begin With, Silver Spring, Maryland).

Barnett, V. (1999). Ranked set sample design for environmental investigations, *Environmental and Ecological Statistics* **6**, 1, pp. 59–74.

Basu, A. and Sarkar, S. (1994). On disparity based goodness-of-fit tests for multinomial models, *Statistics & Probability Letters* **19**, 4, pp. 307–312.

Bateman, G. (1948). On the power function of the longest run as a test for randomness in a sequence of alternatives, *Biometrika* **35**, 1/2, pp. 97–112.

Benjamini, Y. and Hochberg, Y. (1995). Controlling the false discovery rate: a practical and powerful approach to multiple testing, *Journal of the royal statistical society. Series B (Methodological)*, pp. 289–300.

Beran, R. (1977). Minimum hellinger distance estimates for parametric models, *The Annals of Statistics*, pp. 445–463.

Berry, A. C. (1941). The accuracy of the gaussian approximation to the sum of independent variates, *Transactions of the american mathematical society* **49**, 1, pp. 122–136.

Bhapkar, V. (1961). A nonparametric test for the problem of several samples, *The Annals of Mathematical Statistics*, pp. 1108–1117.

Bhapkar, V. and Deshpande, J. V. (1968). Some nonparametric tests for multisample problems, *Technometrics* **10**, 3, pp. 578–585.

Billingsley, P. (1995). *Probability and measure* (A Wiley-Interscience Publication, John Wiley).

Blum, J., Weiss, L., *et al.* (1957). Consistency of certain two-sample tests, *The Annals of Mathematical Statistics* **28**, 1, pp. 242–246.

Bohn, L. L. and Wolfe, D. A. (1992). Nonparametric two-sample procedures for ranked-set samples data, *Journal of the American Statistical Association* **87**, 418, pp. 552–561.

Casella, G. and Berger, R. L. (2002). *Statistical inference*, Vol. 2 (Duxbury Pacific Grove, CA).

Chatterjee, A. and Lahiri, S. (2011). Strong consistency of lasso estimators, *Sankhya A* **73**, 1, pp. 55–78.

Chen, Z., Bai, Z., and Sinha, B. (2003). *Ranked set sampling: theory and applications*, Vol. 176 (Springer Science & Business Media).

Cheng, N. and Yuan, T. (2013). Nonparametric bayesian lifetime data analysis using dirichlet process lognormal mixture model, *Naval Research Logistics (NRL)* **60**, 3, pp. 208–221.

Chernoff, H., Lehmann, E., *et al.* (1954). The use of maximum likelihood estimates in χ^2 tests for goodness of fit, *The Annals of Mathematical Statistics* **25**, 3, pp. 579–586.

Chernoff, H. and Savage, I. R. (1958). Asymptotic normality and efficiency of certain nonparametric test statistics, *The Annals of Mathematical Statistics*, pp. 972–994.

Chow, Y. S. and Teicher, H. (1997). *Encyclopaedia of Mathematical Sciences: Independence, Interchangeability, Martingales. Probability Theory* (Springer).

Conover, W. J. and Conover, W. J. (1980). *Practical nonparametric statistics* (Wiley New York).

Cox, D. R. (1972). Regression models and life tables (with discussion), *Journal of the Royal Statistical Society* **34**, pp. 187–220.

Cramér, H. (1974). *Mathematical Methods of Statistics* (Princeton University-Press, Princeton, NJ).

Cressie, N. and Read, T. R. (1984). Multinomial goodness-of-fit tests, *Journal of the Royal Statistical Society. Series B (Methodological)*, pp. 440–464.

Critchlow, D. E. and Fligner, M. A. (1991). On distribution-free multiple comparisons in the one-way analysis of variance, *Communications in statistics-theory and methods* **20**, 1, pp. 127–139.

d'Agostino, R. B. (1971). An omnibus test of normality for moderate and large size samples, *Biometrika* **58**, 2, pp. 341–348.

David, F. (1947). A power function for tests of randomness in a sequence of alternatives, *Biometrika* **34**, 3/4, pp. 335–339.

David, H. A. and Nagaraja, H. N. (1981). *Order statistics* (Wiley Online Library).

Davison, A. C. and Hinkley, D. V. (1997). *Bootstrap methods and their application*, Vol. 1 (Cambridge university press).

Deheuvels, P. (1979). La fonction de dépendance empirique et ses propriétés. un test non paramétrique d'indépendance, *Acad. Roy. Belg. Bull. Cl. Sci.(5)* **65**, 6, pp. 274–292.

Deshpande, J. V. (1965). A nonparametric test based on u-statistics for the problem of several samples, *J. Indian Statist. Assoc* **3**, 1, pp. 20–29.

Deshpande, J. V. (1972). Linear ordered rank tests which are asymptotically efficient for the two-sample problem, *Journal of the Royal Statistical Society. Series B (Methodological)*, pp. 364–370.

Deshpande, J. V. (1983). A class of tests for exponentiality against increasing failure rate average alternatives, *Biometrika* **70**, 2, pp. 514–518.

Deshpande, J. V. and Kusum, K. (1984). A test for the nonparametric two sample scale problem, *Australian Journal of Statistics* **26**, 1, pp. 16–24.

Deshpande, J. V. and Purohit, S. G. (2016). *Lifetime Data: Statistical Models and Methods, (2nd edition)* (World Scientific).

Dewan, I., Deshpande, J. V., and Kulathinal, S. B. (2004). On testing dependence between time to failure and cause of failure via conditional probabilities, *Scandinavian Journal of Statistics* **31**, 1, pp. 79–91.

Dey, J., Erickson, R., and Ramamoorthi, R. (2003). Some aspects of neutral to right priors, *International statistical review* **71**, 2, pp. 383–401.

Doksum, K. (1974). Tailfree and neutral random probabilities and their posterior distributions, *The Annals of Probability*, pp. 183–201.

Donsker, M. D. (1952). Justification and extension of doob's heuristic approach to the kolmogorov-smirnov theorems, *The Annals of mathematical statistics* **23**, 2, pp. 277–281.

Doob, J. L. *et al.* (1949). Heuristic approach to the kolmogorov-smirnov theorems, *The Annals of Mathematical Statistics* **20**, 3, pp. 393–403.

Dudley, R. M. (2014). *Uniform central limit theorems* (Cambridge Univ Press, 2nd edition).

Dunnett, C. W. (1955). A multiple comparison procedure for comparing several treatments with a control, *Journal of the American Statistical Association* **50**, 272, pp. 1096–1121.

Dwass, M. (1960). Some k-sample rank-order tests, *In I. Olkin, S. G. Ghurye, H. Hoeffding, W. G. Madow, and H. B. Mann (Eds), Contributions to probability and statistics*, pp. 198–202.

Efron, B. (1979). Bootstrap methods: Another look at the jackknife, *Ann. Statist.* **7**, 1, pp. 1–26.

Efron, B., Hastie, T., Johnstone, I., and Tibshirani, R. (2004). Least angle regression, *The Annals of statistics* **32**, 2, pp. 407–499.

Escobar, M. D. (1994). Estimating normal means with a dirichlet process prior, *Journal of the American Statistical Association* **89**, 425, pp. 268–277.

Escobar, M. D. and West, M. (1995). Bayesian density estimation and inference using mixtures, *Journal of the american statistical association* **90**, 430, pp. 577–588.

Esseen, C.-G. (1945). Fourier analysis of distribution functions. a mathematical study of the laplace-gaussian law, *Acta Mathematica* **77**, 1, pp. 1–125.

Feller, W. (1968). *An introduction to probability theory and its applications: volume I*, Vol. 3 (John Wiley & Sons London-New York-Sydney-Toronto).

Feller, W. (1971). *William. An introduction to probability theory and its applications. Vol. II* (John Wiley & Sons, New York).

Ferguson, T. S. (1973). A bayesian analysis of some nonparametric problems, *The annals of statistics*, pp. 209–230.

Ferguson, T. S. (1974). Prior distributions on spaces of probability measures, *The annals of statistics*, pp. 615–629.

Ferguson, T. S. (1983). Bayesian density estimation by mixtures of normal distributions, *Recent advances in statistics* **24**, 1983, pp. 287–302.

Fisher, R. A. (1924). The conditions under which χ^2 measures the discrepancey between observation and hypothesis.

Fisz, M. (1963). *Probability Theory and Mathematical Statistics* (John Wiley & Sons Inc., New York).

Fleming, T. and Harrington, D. (1991). Counting processes and survival analysis (John Wiley & Sons Inc., New York).

Fligner, M. and Wolfe, D. (1982). Distribution-free tests for comparing several treatments with a control, *Statistica Neerlandica* **36**, 3, pp. 119–127.

Fraser, D. A. S. (1957). Nonparametric methods in statistics (Wiley, New York).

Gail, M. H. and Green, S. B. (1976). A generalization of the one-sided two-sample kolmogorov-smirnov statistic for evaluating diagnostic tests, *Biometrics*, pp. 561–570.

Genest, C., Quessy, J.-F., Rémillard, B., et al. (2007). Asymptotic local efficiency of cramér–von mises tests for multivariate independence, *The Annals of Statistics* **35**, 1, pp. 166–191.

Genest, C. and Rémillard, B. (2004). Test of independence and randomness based on the empirical copula process, *Test* **13**, 2, pp. 335–369.

Ghosh, J. and Ramamoorthi, R. (2003). Bayesian nonparametrics, *Springer Series in Statistics*.

Gibbons, J. D. and Chakraborti, S. (2011). *Nonparametric statistical inference (Fifth Edition)* (Springer).

Govindarajulu, Z. (2007). *Nonparametric inference* (World Scientific).

Govindarajulu, Z., Le Cam, L., Raghavachari, M., et al. (1967). *Generalizations of theorems of Chernoff and Savage on the asymptotic normality of test statistics* (The Regents of the University of California).

Gumbel, E. (1958). Statistics of extremes (Columbia Univ. press, New York).

Hájek, Z. and Šidák, J. (1967). *Theory of rank tests* (Academia, Prague & Academic Press, New York-London).

Hall, W. J. and Wellner, J. A. (1980). Confidence bands for a survival curve from censored data, *Biometrika* **67**, 1, pp. 133–143.

Hannan, E. (1956). The asymptotic powers of certain tests based on multiple correlations, *Journal of the Royal Statistical Society. Series B (Methodological)*, pp. 227–233.

Hettmansperger, T. P. (1995). The ranked-set sample sign test, *Journal of Nonparametric Statistics* **4**, 3, pp. 263–270.

Hjort, N. L. (1990). Nonparametric bayes estimators based on beta processes in models for life history data, *The Annals of Statistics*, pp. 1259–1294.

Hochberg, Y. and Tamhane, A. C. (2009). *Multiple comparison procedures* (Wiley).

Hodges Jr, J. L. and Lehmann, E. L. (1963). Estimates of location based on rank tests, *The Annals of Mathematical Statistics*, pp. 598–611.

Hoeffding, W. (1940). *Massstabinvariante korrelationstheorie* (Teubner).

Hoeffding, W. (1948). A class of statistics with asymptotically normal distribution, *The annals of mathematical statistics*, pp. 293–325.

Hoeffding, W. (1951). 'optimum' nonparametric tests, *Proc. Second Berkeley Symp. Math. Statist. Probab.*, pp. 83–92.

Hoerl, A. E., Kannard, R. W., and Baldwin, K. F. (1975). Ridge regression: some simulations, *Communications in Statistics-Theory and Methods* **4**, 2, pp. 105–123.

Hollander, M., Proschan, F., *et al.* (1972). Testing whether new is better than used, *The Annals of Mathematical Statistics* **43**, 4, pp. 1136–1146.

Hollander, M., Wolfe, D. A., and Chicken, E. (2013). *Nonparametric statistical methods (Third Edition)* (John Wiley & Sons).

Hundal, P. (1969). Knowledge of performance as an incentive in repetitive industrial work. *Journal of Applied Psychology* **53**, 3pt. 1, pp 224–226.

James, G., Witten, D., Hastie, T., and Tibshirani, R. (2013). *An introduction to statistical learning*, Vol. 6 (Springer).

Jonckheere, A. R. (1954). A distribution-free k-sample test against ordered alternatives, *Biometrika* **41**, 1/2, pp. 133–145.

Kaplan, E. L. and Meier, P. (1958). Nonparametric estimation from incomplete observations, *Journal of the American statistical association* **53**, 282, pp. 457–481.

Karr, A. (1991). *Point processes and their statistical inference*, Vol. 7 (CRC press).

Kiefer, J. and Wolfowitz, J. (1956). Consistency of the maximum likelihood estimator in the presence of infinitely many incidental parameters, *The Annals of Mathematical Statistics*, pp. 887–906.

Kim, Y. and Lee, J. (2001). On posterior consistency of survival models, *Annals of Statistics*, pp. 666–686.

Klein, J. P. and Moeschberger, M. L. (2005). *Survival analysis: techniques for censored and truncated data* (Springer Science & Business Media).

Knight, K. and Fu, W. (2000). Asymptotics for lasso-type estimators, *Annals of Statistics*, pp. 1356–1378.

Kochar, S. and Deshpande, J. V. (1988). Nonparametric tests for independence in bivariate reliability models, *Proceedings of the International Conference on Advances in Multivariate Statistical Analysis, edited by J. K. Ghosh and S. DasGupta*, pp. 203–217.

Kolmogorov, A. N. (1933). Sulla determinazione empirica di una legge di distribuzione, *Inst. Ital. Attuari, Giorn.* **4**, pp. 83–91.

Koti, K. M. and Jogesh Babu, G. (1996). Sign test for ranked-set sampling, *Communications in Statistics-Theory and Methods* **25**, 7, pp. 1617–1630.

Kottas, A. (2006). Nonparametric bayesian survival analysis using mixtures of weibull distributions, *Journal of Statistical Planning and Inference* **136**, 3, pp. 578–596.

Lai, C. and Xie, M. (2000). A new family of positive quadrant dependent bivariate distributions, *Statistics & probability letters* **46**, 4, pp. 359–364.

Lavine, M. (1992). Some aspects of polya tree distributions for statistical modelling, *The Annals of Statistics*, pp. 1222–1235.

Lavine, M. (1994). More aspects of polya tree distributions for statistical modelling, *The Annals of Statistics*, pp. 1161–1176.

Lehmann, E. L. (1951). Consistency and unbiasedness of certain nonparametric tests, *The Annals of Mathematical Statistics*, pp. 165–179.

Lehmann, E. L. (1966). Some concepts of dependence, *The Annals of Mathematical Statistics*, pp. 1137–1153.

Lehmann, E. L. (1991). Theory of point estimation, *Wadsworth and Brooks/Cole, Pacific Grove, CA*.

Lehmann, E. L. (1975). Nonparametrics: statistical methods based on ranks. With the special assistance of H. J. M. d'Abrera. Holden-Day Series in Probability and Statistics. Holden-Day, Inc., San Francisco, Calif.

Lehmann, E. L. and Romano, J. P. (2006). *Testing statistical hypotheses* (Springer Science & Business Media).

Lieblein, J. and Zelen, M. (1956). Statistical investigation of the fatigue life of deep-groove ball bearings, *Journal of Research of the National Bureau of Standards* **57**, 5, pp. 273–316.

Lo, A. Y. *et al.* (1984). On a class of bayesian nonparametric estimates: I. density estimates, *The annals of statistics* **12**, 1, pp. 351–357.

Loève, M. (1963). Probability theory, *Graduate texts in mathematics* **45**.

Mann, H. B. and Whitney, D. R. (1947). On a test of whether one of two random variables is stochastically larger than the other, *The annals of mathematical statistics*, pp. 50–60.

Mauldin, R. D., Sudderth, W. D., and Williams, S. (1992). Polya trees and random distributions, *The Annals of Statistics*, pp. 1203–1221.

McIntyre, G. (1952). A method for unbiased selective sampling, using ranked sets, *Crop and Pasture Science* **3**, 4, pp. 385–390.

Miller, R. G. (1981). *Simultaneous statistical inference* (Springer).

Moore, D. *et al.* (1968). An elementary proof of asymptotic normality of linear functions of order statistics, *The Annals of Mathematical Statistics* **39**, 1, pp. 263–265.

Müller, P., Erkanli, A., and West, M. (1996). Bayesian curve fitting using multivariate normal mixtures, *Biometrika* **83**, 1, pp. 67–79.

Naik-Nimbalkar, U. and Rajarshi, M. (1997). Empirical likelihood ratio test for equality of k medians in censored data, *Statistics & probability letters* **34**, 3, pp. 267–273.

Nelsen, R. B. (1992). On measures of association as measures of positive dependence, *Statistics & probability letters* **14**, 4, pp. 269–274.

Nelson, W. (1969). Hazard plotting for incomplete failure data, *Journal of Quality Technology* **1**, 1, pp. 27–52.

Noether, G. E. (1955). On a theorem of Pitman, *The Annals of Mathematical Statistics*, **26**, pp. 64–68.

Owen, A. B. (1998). *Empirical Likelihood* (Wiley Online Library).

Öztürk, Ö. and Deshpande, J. V. (2004). A new nonparametric test using ranked set data for a two-sample scale problem, *Sankhyā: The Indian Journal of Statistics*, pp. 513–527.

Öztürk, Ö. and Deshpande, J. V. (2006). Ranked-set sample nonparametric quantile confidence intervals, *Journal of Statistical Planning and Inference* **136**, 3, pp. 570–577.

Öztürk, Ö. and Wolfe, D. A. (2000). An improved ranked set two-sample mann-whitney-wilcoxon test, *Canadian Journal of Statistics* **28**, 1, pp. 123–135.

Patil, G., Sinha, A., and Taillie, C. (1994). 5 ranked set sampling, *Handbook of statistics* **12**, pp. 167–200.

Pearson, K. (1900). X. on the criterion that a given system of deviations from the probable in the case of a correlated system of variables is such that it can be reasonably supposed to have arisen from random sampling, *The London, Edinburgh, and Dublin Philosophical Magazine and Journal of Science* **50**, 302, pp. 157–175.

Petrone, S. and Veronese, P. (2002). Non parametric mixture priors based on an exponential random scheme, *Statistical Methods and Applications* **11**, 1, pp. 1–20.

Puri, M. L., Sen, P. K., *et al.* (1971). *Nonparametric methods in multivariate analysis* (Wiley).

Pyke, R. and Shorack, G. R. (1968). Weak convergence of a two-sample empirical process and a new approach to chernoff-savage theorems, *The Annals of Mathematical Statistics*, pp. 755–771.

Randles, R. H. and Wolfe, D. A. (1979). *Introduction to the theory of nonparametric statistics*, Vol. 1 (Wiley New York).

Read, T. R. and Cressie, N. A. (2012). *Goodness-of-fit statistics for discrete multivariate data* (Springer Science & Business Media).

Rohatgi, V. K. and Saleh, A. M. E. (2015). *An introduction to probability and statistics* (John Wiley & Sons).

Rothe, G. (1981). Some properties of the asymptotic relative pitman efficiency, *The Annals of Statistics*, pp. 663–669.

Savage, I. R. (1956). Contributions to the theory of rank order statistics-the two-sample case, *The Annals of Mathematical Statistics* **27**, 3, pp. 590–615.

Scheffe, H. (1953). A method for judging all contrasts in the analysis of variance, *Biometrika* **40**, 1-2, pp. 87–110.

Sen, P. K. (1963). On the estimation of relative potency in dilution (-direct) assays by distribution-free methods, *Biometrics*, pp. 532–552.

Sethuraman, J. (1994). A constructive definition of dirichlet priors, *Statistica sinica*, pp. 639–650.

Shapiro, S. S. and Francia, R. (1972). An approximate analysis of variance test for normality, *Journal of the American Statistical Association* **67**, 337, pp. 215–216.

Shin, D. W., Basu, A., and Sarkar, S. (1995). Comparisons of the blended weight hellinger distance based goodness-of-fit test statistics, *Sankhyā: The Indian Journal of Statistics, Series B* **57**, pp. 365–376.

Shorack, G. R. and Wellner, J. A. (1986). Empirical processes with applications to statistics wiley series in probability and mathematical statistics: Probability and mathematical statistics, *New York*.

Siegel, S. (1956). *Nonparametric statistics for the behavioral sciences* (McGraw-hill).

Silva, J. L. D. (2014). Some strong consistency results in stochastic regression, *Journal of Multivariate Analysis* **129**, pp. 220–226.

Silva, J. L. D., Mexia, J. T., and Ramos, L. P. (2015). On the strong consistency of ridge estimates, *Communications in Statistics-Theory and Methods* **44**, 3, pp. 617–626.

Singh, K. (1984). Asymptotic comparison of tests — a review, *Handbook of Statistics* **4**, pp. 173–184.

Smirnov, N. (1948). Table for estimating the goodness of fit of empirical distributions, *The annals of mathematical statistics* **19**, 2, pp. 279–281.

Smirnov, N. V. (1939). On the estimation of the discrepancy between empirical curves of distribution for two independent samples, *Bull. Math. Univ. Moscou* **2**, 2.

Sprent, P. and Smeeton, N. C. (2016). *Applied nonparametric statistical methods (Fourth edition)* (CRC Press).

Spurrier, J. D. (2006). Additional tables for Steel–Dwass–Critchlow–Fligner distribution-free multiple comparisons of three treatments, *Communications in Statistics–Simulation and Computation®* **35**, 2, pp. 441–446.

Steel, R. G. (1960). A rank sum test for comparing all pairs of treatments, *Technometrics* **2**, 2, pp. 197–207.

Steel, R. G. (1961). Some rank sum multiple comparisons tests, *Biometrics*, pp. 539–552.

Stigler, S. M. (1974). Linear functions of order statistics with smooth weight functions, *The Annals of Statistics*, pp. 676–693.

Sukhatme, B. V. (1958). A two sample distribution free test for comparing variances, *Biometrika* **45**, 3/4, pp. 544–548.

Tchen, A. H. (1980). Inequalities for distributions with given marginals, *The Annals of Probability*, pp. 814–827.

Terry, M. E. (1952). Some rank order tests which are most powerful against specific parametric alternatives, *The Annals of Mathematical Statistics*, pp. 346–366.

Tibshirani, R. (1996). Regression shrinkage and selection via the lasso, *Journal of the Royal Statistical Society. Series B (Methodological)*, pp. 267–288.

Tukey, J. W. (1991). The philosophy of multiple comparisons, *Statistical Science*, pp. 100–116.

Wald, A. and Wolfowitz, J. (1940). On a test whether two samples are from the same population, *The Annals of Mathematical Statistics* **11**, 2, pp. 147–162.

Wasserman, L. (2007). *All of Nonparametric Statistics* (Springer, New York).

Wilcoxon, F. (1945). Individual comparisons by ranking methods, *Biometrics bulletin* **1**, 6, pp. 80–83.

Wilks, S. S. (1962). *Mathematical statistics* (New York, John Wiley and Sons).

Wolfe, D. A. (2004). Ranked set sampling: an approach to more efficient data collection, *Statistical Science*, pp. 636–643.

INDEX